REPORT OF A WORKSHOP ON THE PEDAGOGICAL ASPECTS OF
COMPUTATIONAL THINKING

Committee for the Workshops on Computational Thinking

Computer Science and Telecommunications Board

Division on Engineering and Physical Sciences

NATIONAL RESEARCH COUNCIL
OF THE NATIONAL ACADEMIES

THE NATIONAL ACADEMIES PRESS
Washington, D.C.
www.nap.edu

THE NATIONAL ACADEMIES PRESS   500 Fifth Street, N.W.   Washington, DC 20001

NOTICE: The project that is the subject of this report was approved by the Governing Board of the National Research Council, whose members are drawn from the councils of the National Academy of Sciences, the National Academy of Engineering, and the Institute of Medicine. The members of the committee responsible for the report were chosen for their special competences and with regard for appropriate balance.

Support for this project was provided by the National Science Foundation under sponsor award number CNS-0831827. Any opinions expressed in this material are those of the authors and do not necessarily reflect the views of the agencies and organizations that provided support for the project.

International Standard Book Number-13:   978-0-309-21474-2
International Standard Book Number-10:   0-309-21474-2

Copies of this report are available from

The National Academies Press
500 Fifth Street, N.W., Lockbox 285
Washington, DC 20055
800/624-6242
202/334-3313 (in the Washington metropolitan area)
http://www.nap.edu

Copyright 2011 by the National Academy of Sciences. All rights reserved.

Printed in the United States of America

THE NATIONAL ACADEMIES
Advisers to the Nation on Science, Engineering, and Medicine

The **National Academy of Sciences** is a private, nonprofit, self-perpetuating society of distinguished scholars engaged in scientific and engineering research, dedicated to the furtherance of science and technology and to their use for the general welfare. Upon the authority of the charter granted to it by the Congress in 1863, the Academy has a mandate that requires it to advise the federal government on scientific and technical matters. Dr. Ralph J. Cicerone is president of the National Academy of Sciences.

The **National Academy of Engineering** was established in 1964, under the charter of the National Academy of Sciences, as a parallel organization of outstanding engineers. It is autonomous in its administration and in the selection of its members, sharing with the National Academy of Sciences the responsibility for advising the federal government. The National Academy of Engineering also sponsors engineering programs aimed at meeting national needs, encourages education and research, and recognizes the superior achievements of engineers. Dr. Charles M. Vest is president of the National Academy of Engineering.

The **Institute of Medicine** was established in 1970 by the National Academy of Sciences to secure the services of eminent members of appropriate professions in the examination of policy matters pertaining to the health of the public. The Institute acts under the responsibility given to the National Academy of Sciences by its congressional charter to be an adviser to the federal government and, upon its own initiative, to identify issues of medical care, research, and education. Dr. Harvey V. Fineberg is president of the Institute of Medicine.

The **National Research Council** was organized by the National Academy of Sciences in 1916 to associate the broad community of science and technology with the Academy's purposes of furthering knowledge and advising the federal government. Functioning in accordance with general policies determined by the Academy, the Council has become the principal operating agency of both the National Academy of Sciences and the National Academy of Engineering in providing services to the government, the public, and the scientific and engineering communities. The Council is administered jointly by both Academies and the Institute of Medicine. Dr. Ralph J. Cicerone and Dr. Charles M. Vest are chair and vice chair, respectively, of the National Research Council.

www.national-academies.org

COMMITTEE FOR THE WORKSHOPS ON COMPUTATIONAL THINKING

MARCIA LINN, University of California, Berkeley, *Chair*
ALFRED V. AHO, Columbia University
M. BRIAN BLAKE, University of Notre Dame
ROBERT CONSTABLE, Cornell University
YASMIN B. KAFAI, University of Pennsylvania
JANET L. KOLODNER, Georgia Institute of Technology
LAWRENCE SNYDER, University of Washington, Seattle
URI WILENSKY, Northwestern University

Staff

HERBERT S. LIN, Study Director and Chief Scientist, CSTB
ENITA A. WILLIAMS, Associate Program Officer
SHENAE BRADLEY, Senior Program Assistant

COMPUTER SCIENCE AND TELECOMMUNICATIONS BOARD

ROBERT F. SPROULL, Oracle, *Chair*
PRITHVIRAJ BANERJEE, Hewlett Packard Company
STEVEN M. BELLOVIN, Columbia University
SEYMOUR E. GOODMAN, Georgia Institute of Technology
JOHN E. KELLY III, IBM Research
JON M. KLEINBERG, Cornell University
ROBERT KRAUT, Carnegie Mellon University
SUSAN LANDAU, Radcliffe Institute for Advanced Study
PETER LEE, Microsoft Corporation
DAVID LIDDLE, US Venture Partners
WILLIAM H. PRESS, University of Texas at Austin
PRABHAKAR RAGHAVAN, Yahoo! Research
DAVID E. SHAW, D.E. Shaw Research
ALFRED Z. SPECTOR, Google, Inc.
JOHN SWAINSON, Silver Lake Partners
PETER SZOLOVITS, Massachusetts Institute of Technology
PETER J. WEINBERGER, Google, Inc.
ERNEST J. WILSON, University of Southern California

JON EISENBERG, Director
RENEE HAWKINS, Financial and Administrative Manager
HERBERT S. LIN, Chief Scientist, CSTB
LYNETTE I. MILLETT, Senior Program Officer
EMILY ANN MEYER, Program Officer
VIRGINIA BACON TALATI, Associate Program Officer
ENITA A. WILLIAMS, Associate Program Officer
SHENAE BRADLEY, Senior Program Assistant
ERIC WHITAKER, Senior Program Assistant

For more information on CSTB, see its website at http://www.cstb.org, write to CSTB, National Research Council, 500 Fifth Street, N.W., Washington, D.C. 20001, call (202) 334-2605, or e-mail the CSTB at cstb@nas.edu.

Preface

In 2008, the Computer and Information Science and Engineering Directorate of the National Science Foundation asked the National Research Council (NRC) to conduct two workshops to explore the nature of computational thinking and its cognitive and educational implications. The first workshop focused on the scope and nature of computational thinking and on articulating what "computational thinking for everyone" might mean. A report of that workshop was released in January 2010.[1] Drawing in part on the proceedings of that workshop, the present report summarizes the second workshop, which was held February 4-5, 2010, in Washington, D.C., and focused on pedagogical considerations for computational thinking.

Although this document was prepared by the Committee for the Workshops on Computational Thinking based on workshop presentations and discussions, it does not reflect consensus views of the committee. Under NRC guidelines for conducting workshops and developing workshop report summaries, workshop activities do not seek consensus and workshop reports (such as the present volume) cannot be said to represent "an NRC view" on the subject at hand. As with the first workshop, this second workshop revealed a plethora of perspectives on ways to approach pedagogy for computational thinking. The two workshops,

---

[1] National Research Council, 2010, *Report of a Workshop on the Scope and Nature of Computational Thinking*, Washington, D.C.: The National Academies Press. Available at http://www.nap.edu/catalog.php?record_id=12840. Last accessed February 7, 2011.

taken together, call attention to the diversity of views on many aspects of computational thinking as well as its definition, and it is the hope of the committee that the present report, which contains a digest of both presentations and discussion, will serve as a vehicle that increases communication on the topic across the community.

The full workshop agenda is provided in Appendix A, and short biographies of the workshop participants are given in Appendix B.

                    Marcia C. Linn, *Chair*
                    Committee for the Workshops on Computational Thinking

Acknowledgment of Reviewers

This report has been reviewed in draft form by individuals chosen for their diverse perspectives and technical expertise, in accordance with procedures approved by the National Research Council's (NRC's) Report Review Committee. The purpose of this independent review is to provide candid and critical comments that will assist the institution in making its published report as sound as possible and to ensure that the report meets institutional standards for objectivity, evidence, and responsiveness to the study charge. The review comments and draft manuscript remain confidential to protect the integrity of the deliberative process. We wish to thank the following individuals for their review of this report:

Christine Cunningham, Museum of Science
Margaret Honey, New York Hall of Science
Peter Szolovits, Massachusetts Institute of Technology
Robert Tinker, The Concord Consortium
Michelle Williams, Michigan State University

Although the reviewers listed above have provided many constructive comments and suggestions, they did not see the final draft of the report before its release. The review of this report was coordinated by Joseph F. Traub, Columbia University. Appointed by the NRC, he was responsible for making certain that an independent examination of this report was carried out in accordance with institutional procedures and that all review comments were carefully considered. Responsibility for the final content of this report rests entirely with the authoring committee and the institution.

Contents

1 INTRODUCTION 1
   1.1 Scope and Approach of This Report, 1
   1.2 Motivating an Examination of Pedagogy, 4
   1.3 Organization of This Report, 5

2 KEY POINTS EXPRESSED BY PRESENTERS AND
   DISCUSSANTS 6
   2.1 Perspectives on Computational Thinking and
      Computational Thinkers, 6
   2.2 Activities of Computational Thinking, 7
   2.3 Contexts for Computational Thinking, 9
      2.3.1 Everyday Life, 10
      2.3.2 Games and Gaming, 10
      2.3.3 Science, 11
      2.3.4 Engineering, 15
      2.3.5 Journalism, 15
      2.3.6 Abstracting Problem Solving from Specific Contexts, 16
   2.4 Pedagogical Environments for Computational Thinking, 17
      2.4.1 Foci for Pedagogical Environments, 17
      2.4.2 Illustrative Pedagogical Environments, 19
   2.5 Developmental Considerations and Computational
      Thinking, 21
      2.5.1 Development of Scientific Thinking in Children, 23
      2.5.2 Possible Progressions, 24

2.6 Assessments for Computational Thinking, 26
2.7 Teachers and Computational Thinking, 27
  2.7.1 Professional Development and Other Needs for Teacher Support, 28
  2.7.2 Teachers and Career Awareness, 29
2.8 Learning Contexts and Computational Thinking, 30
  2.8.1 Aligning with Standards, 30
  2.8.2 Out-of-School Computational Thinking, 31
2.9 Research and Unanswered Questions Regarding Computational Thinking, 33
  2.9.1 The Importance of a Process for Defining Computational Thinking, 33
  2.9.2 The Role of Technology, 34
  2.9.3 The Need for Interoperability, 34
  2.9.4 The Need for a Career Framework, 35

3 COMMITTEE MEMBER PERSPECTIVES 36
  3.1 Alfred Aho, 36
  3.2 Uri Wilensky, 39
  3.3 Yasmin Kafai, 45
  3.4 Marcia Linn, 47
  3.5 Larry Snyder, 50
  3.6 Janet Kolodner, 52
  3.7 Brian Blake, 63

4 SUMMARIES OF INDIVIDUAL PRESENTATIONS 65
  4.1 Computational Thinking and Scientific Visualization, 65
    4.1.1 Questions Addressed, 65
    4.1.2 Robert Tinker, Concord Consortium, 66
    4.1.3 Mitch Resnick, Massachusetts Institute of Technology, 67
    4.1.4 John Jungck, Beloit College, BioQUEST, 69
    4.1.5 Idit Caperton, World Wide Workshop, Globaloria, 72
  4.2 Computational Thinking and Technology, 75
    4.2.1 Questions Addressed, 75
    4.2.2 Robert Panoff, Shodor Education Foundation, 75
    4.2.3 Stephen Uzzo, New York Hall of Science, Museum Studies, 78
    4.2.4 Jill Denner, Education, Training, Research Associates, 81
    4.2.5 Lou Gross, National Institute for Mathematical and Biological Synthesis, 83
  4.3 Computational Thinking in Engineering and Computer Science, 86
    4.3.1 Questions Addressed, 86

4.3.2 Christine Cunningham, Museum of Science, Engineering is Elementary Project, 87
4.3.3 Taylor Martin, University of Texas at Austin, 91
4.3.4 Ursula Wolz, College of New Jersey, 92
4.3.5 Peter Henderson, Butler University, 95
4.4 Teaching and Learning Computational Thinking, 97
4.4.1 Questions Addressed, 97
4.4.2 Deanna Kuhn, Columbia University, 97
4.4.3 Matthew Stone, Rutgers University, 99
4.4.4 Jim Slotta, University of Toronto, Ontario Institute for Studies in Education, 101
4.4.5 Joyce Malyn-Smith, Education Development Center, Inc., ITEST Learning Resource Center, 104
4.4.6 Jan Cuny, National Science Foundation, CS 10K Project, 108
4.5 Educating the Educators, 110
4.5.1 Questions Addressed, 110
4.5.2 Michelle Williams, Michigan State University, 111
4.5.3 Walter Allan and Jeri Erickson, Foundation for Blood Research, EcoScienceWorks Project, 115
4.5.4 Danny Edelson, National Geographic Society, 117
4.6 Measuring Outcomes (for Evaluation) and Collecting Feedback (for Assessment), 120
4.6.1 Questions Addressed, 120
4.6.2 Paulo Blikstein, Stanford University, 121
4.6.3 Christina Schwarz, Michigan State University, 123
4.6.4 Mike Clancy, University of California, Berkeley, 126
4.6.5 Derek Briggs, University of Colorado, Boulder, 128
4.6.6 Cathy Lachapelle, Museum of Science, Engineering is Elementary Project, 130

5 CONCLUSION 133

APPENDIXES

A Workshop Agenda 137
B Short Biographies of Committee Members, Workshop Participants, and Staff 143

1

Introduction

1.1 SCOPE AND APPROACH OF THIS REPORT

This report summarizes the second of two workshops on computational thinking, which was held February 4-5, 2010, in Washington, D.C., under the auspices of the National Research Council's (NRC's) Committee for the Workshops on Computational Thinking.[1] This second workshop was structured to gather pedagogical inputs and insights from educators who have addressed computational thinking in their work with K-12 teachers and students.

Questions posed to participants in the second workshop included the following:

- What are the relevant lessons learned and best practices for improving computational thinking in K-12 education?
- What are some examples of computational thinking and how, if at all, does computational thinking vary by discipline at the K-12 level?
- What exposures and experiences contribute to developing computational thinking in the disciplines? What are some innovative environments for teaching computational thinking?

---

[1] The first workshop, held February 19-20, 2009, is summarized in National Research Council, 2010, *Report of a Workshop on the Scope and Nature of Computational Thinking*, Washington, D.C.: The National Academies Press. Available at http://www.nap.edu/catalog.php?record_id=12840. Last accessed February 7, 2011.

- Is there a progression of computational thinking concepts in K-12 education? What are some criteria by which to order such a progression?
- How should professional development efforts and classroom support be adapted to the varying experience levels of teachers such as pre-service, inducted, and in-service levels? What tools are available to support teachers as they teach computational thinking?
- How does computational thinking education connect with other subjects? Should computational thinking be integrated in other subjects taught in the classroom?
- How can learning of computational thinking be assessed? How should we measure the success of efforts to teach computational thinking?

This workshop was structured to illuminate different approaches to the teaching of computational thinking. Participants often clarified their own interpretations of computational thinking in relation to the discussion in the first workshop report.

To improve readability and to promote understanding, background material on some of the topics and ideas raised is interspersed in this workshop report. This workshop report also includes some of the material discussed in the first workshop that related to pedagogy and how best to expose students to the ideas of computational thinking but that was not addressed in the first workshop report.

The second workshop was deliberately organized to include individuals with a broad range of perspectives. For this reason and because some of the discussion amounted to brainstorming, this workshop summary may contain internal inconsistencies that reflect the wide range of views offered by workshop participants. In keeping with its purpose of exploring the topic, this workshop summary does not contain findings or recommendations.

The reader is cautioned that the workshop was not intended to result in a consensus regarding the scope and nature of computational thinking. As was true in the first workshop, participants in the second workshop expressed a host of different views about the scope and nature of computational thinking. As stated in the first report:

> Even though workshop participants generally did not explicitly disagree with views of computational thinking that were not identical to their own, almost every participant held his or her own perspective on computational thinking that placed greater emphasis on particular aspects or characteristics of importance to that individual.[2]

---

[2] National Research Council, 2010, *Report of a Workshop on the Scope and Nature of Computational Thinking*, Washington, D.C.: The National Academies Press, p. 59. Available at http://www.nap.edu/catalog.php?record_id=12840. Last accessed February 7, 2011.

The first report raised the possibility that given this variation in individual perspectives, one possibility concerning the structure of computational thinking is that computational thinking should be regarded simply as "the union of these different views—a laundry list of different characteristics" (p. 59). Noting that most participants in the first workshop would have found this view deeply unsatisfying, the first report pointed out the value of addressing a number of questions that emerged from the first workshop, including the following:

- What is the core of computational thinking?
- What are the elements of computational thinking?
- What is the sequence or trajectory of development of computational thinking?
- Does computational thinking vary by discipline?

Similar questions regarding the structure and content of computational thinking were raised in the second workshop as well. For example, Joyce Malyn-Smith of the Education Development Center, Inc., said that adopting a consistent definition of computational thinking is necessary because people see computational thinking through only their own lenses—and efforts to advocate for computational thinking in the curriculum will not be credible in the absence of a consensus about its structure and content. Al Aho from Columbia University acknowledged the community's need for a common definition of computational thinking, which was inherently difficult given the rapidly changing world to which computational thinking is often applied. Any static definition of computational thinking likely would be obsolete 10 or 20 years from now, he argued, and thus, "The real challenge for the entire community is to define computational thinking and also to keep it current."

Recognizing that there is no easy-to-summarize definition of computational thinking, the first report noted the view of many computer scientists that computational thinking is a fundamental analytical skill that "everyone, not just computer scientists, can use to help solve problems, design systems, and understand human behavior. [As such,] computational thinking is comparable . . . to the mathematical, linguistic, and logical reasoning . . . taught to all children" (p. 3).

The first report also noted that as usually construed, computational thinking includes "a broad range of mental tools and concepts from computer science that help people solve problems, design systems, understand human behavior, and engage computers to assist in automating a wide range of intellectual processes" (p. 3). The report went on to say that

> Computational thinking might include reformulation of difficult problems by reduction and transformation; approximate solutions; paral-

lel processing; type checking and model checking as generalizations of dimensional analysis; problem abstraction and decomposition; problem representation; modularization; error prevention, testing, debugging, recovery, and correction; damage containment; simulation; heuristic reasoning; planning, learning, and scheduling in the presence of uncertainty; search strategies; analysis of the computational complexity of algorithms and processes; and balancing computational costs against other design criteria. Concepts from computer science such as algorithm, process, state machine, task specification, formal correctness of solutions, machine learning, recursion, pipelining, and optimization also find broad applicability. (p. 3)

Participants in the first workshop discussed computational thinking as a range of concepts, applications, tools, and skill sets; as a language for expression; as the automation of abstractions; and as a cognitive tool. They further commented on how it is related to thinking skills and habits of mind associated with mathematics and engineering, and how various aspects of computational thinking (problem solving/debugging, testing, data mining and information retrieval, concurrency and parallelism, and modeling) are applicable to various disciplines. Many of these ideas were reflected in the second workshop as well.

1.2 MOTIVATING AN EXAMINATION OF PEDAGOGY

Participants in the first workshop offered a number of reasons for promulgating computational thinking skills broadly in the K-12 curriculum. These included succeeding in a technological society, increasing interest in the information technology professions, maintaining and enhancing U.S. economic competitiveness, supporting inquiry in other disciplines, and enabling personal empowerment.

To launch the second workshop, Jeannette Wing, then assistant director of NSF's Computer and Information Science and Engineering Directorate, discussed her goal for a workshop on pedagogy. She argued that an application of the science of learning research in designing grade- and age-appropriate curricula for computational thinking is necessary to maximize its impact on and significance for K-12 students.

Wing pointed to mathematics as a field that has been successful in developing learning progressions that have a solid foundation in research on the human brain and how it learns mathematical concepts. She noted that humans have an innate understanding of relative quantities—they have the ability in many situations to distinguish between larger and smaller quantities at a very early age. This level of recognition suggests that mathematical activities involving concepts of "greater than" and "less than" might be appropriate for very young students. Symbolic representations require different kinds of mental processing, and such processing is usually not possible until later in development, suggesting that activities involving symbolic representation would better be undertaken later in the K-12 sequence.

As a point of contrast, Wing pointed to the way that computer science is typically taught. What is often taught in computer science to middle school and high school students, said Wing, reflects a relatively casual approach and is a minimally modified version of what is taught to undergraduates. What makes the approach casual is that it is done without a deep appreciation for how students learn at different ages. She noted that there is not a body of grounded and research-based knowledge about how the various aspects of computational thinking or computing map on to brain development. She went on to point out that despite this lack of knowledge, many people believe that some of the abstract concepts of computational thinking cannot be taught before students enter the eighth grade, because of a common assumption that only at that age are students able to learn abstract concepts. There are many such assumptions, she said, that must be evaluated in light of serious research about learning, research that has not yet been done with reference to computing or computational thinking.

1.3 ORGANIZATION OF THIS REPORT

Most of the workshop was devoted to describing and discussing various approaches to the teaching of computational thinking.

This report is organized as follows. Chapter 2 provides what the committee believes to be the key points raised by workshop participants—that is, the committee extracted from the various presentations and discussion sessions a number of key points that in its judgment speak most closely to the teaching of computational thinking. Chapter 3 presents individual committee members' personal synthesis of points made in the respective panel sessions that they moderated. Chapter 4 contains summaries of individual presentations by workshop participants, which often elaborate in more detail examples described in Chapter 2. Depending on the depth and degree of context in which the reader is interested, the committee encourages reading back and forth between these different levels of summary.

Although workshop participants did not agree explicitly on a definition of computational thinking, the examples they provided during this workshop are valuable as indicators of ways that people see the intersection of computation, disciplinary knowledge, and algorithms. Other examples identify what the participants saw as issues and problems when trying to introduce computational thinking into school and non-school pedagogical contexts.

2

Key Points Expressed by Presenters and Discussants

2.1 PERSPECTIVES ON COMPUTATIONAL THINKING AND COMPUTATIONAL THINKERS

Workshop participants extended the discussion started at the first workshop concerning the nature of computational thinking and computational thinkers. In offering one perspective, Peter Henderson, formerly chair of the Department of Computer Science and Software Engineering at Butler University, described computational thinking as generalized problem solving with constraints. He argued that almost every problem-solving activity involves computation of some kind. For Henderson, a toolsmith metaphor is a convenient means for characterizing the elements of computer science and also computational thinking—computer science offers sophisticated tools that strengthen problem solving. Henderson illustrated his point using an example from Thomas the Tank Engine—a series for preschool students. In one situation, Thomas is pulling two cars, one red and one green. They are on a track with a siding (connected on both sides), and the problem is to reverse the order of the two cars. This problem engages students in using a computational algorithm at a very early age.

Matthew Stone, a computational linguist at the Rutgers University's Department of Computer Science and Center for Cognitive Science, argued that core ideas of computational thinking arise in many domains independent of computer technology. Stone pointed out the universality of computational thinking in the context of Jacquard looms that control the weaving of patterns, algorithmic approaches to choosing the grocery checkout line with the shortest wait time, and representation of correspondences between symbols and the physical world such as between online banking and money.

Robert M. Panoff, founder and executive director of the Shodor Education Foundation, Inc., illustrated the generality of computational thinking by identifying three fundamental ideas that ground computational thinking:

- What you have now is what you had before plus what has changed. That is, $X_{new} = X_{old}$ + change in X.
- I am the average of my neighbors; that is, add up a bunch of numbers and divide by the number of numbers. This is the essence of solving Laplace's equation.
- When two entities interact with each other, one of the entities acquires with some probability a property that the other entity already had. For example, if the two entities are people and the property is a wallet, there is some probability of a crime—an example found in criminology. If the entities are trees and the property is being on fire, there is some probability that the tree not on fire will become a tree that is on fire—an example from forest management. If the entities are particles and the property is momentum, there is some probability that one particle will acquire some of the momentum of the other particle—an example often found in physics.

Ursula Wolz, associate professor of computer science and interactive multimedia at the College of New Jersey, noted that concepts of computational thinking permeate journalism. The similarities stem from the reliance of both fields on language. Languages can be natural as found in journalism or formal as found in computer science. Both formal and informal languages involve access to information, aggregation of data, and synthesis of information. Concepts of reliability, privacy, accuracy, and logical consistency are essential to both formal and informal languages. Both involve knowledge representation (e.g., determining the appropriate granularity for reporting a story or taking data) and abstraction from cases.

2.2 ACTIVITIES OF COMPUTATIONAL THINKING

Workshop participants extended the discussion of activities associated with computational thinking that had been initiated at the first workshop. During the second workshop participants focused on educationally relevant activities.

Robert Tinker, founder of the Concord Consortium, argued that the core of computational thinking is to break big problems into smaller problems that lend themselves to efficient, automated solutions. This approach can be implemented using realistic situations as well as visualizations. Tinker advocated introducing computational thinking in science for several reasons. Modern science regularly relies on computational models that are based on scientific principles and are illustrated using visualizations. For example, scientists explore visualizations of new proteins or of new theoretical accounts of tectonic plate movements.

Consulting scientist Walter Allan and outreach education coordinator Jeri Erickson, at ScienceWorks for ME of the Foundation for Blood Research, echoed this point. They argued that the ability to construct rules to specify the behavior of an agent is important to computational thinking. These rules might implement a scientific principle.

Tinker said that he favors exposing students to computational thinking in the context of scientific models and visualizations that depict phenomena in a realistic time sequence. Examples include visualizations of chemical interactions using software such as Molecular Workbench;[1] of force and motion; and of plate tectonics. Students can interact with these visualizations, explore their behavior and limitations, and learn about the science represented in the model. This approach is consistent with the way scientists learn from visualizations and also resonates with the ways that scientists explore the natural world using the scientific method.

Mitch Resnick, professor of learning research at the MIT Media Lab, said that the ability to use computational media to create, build, and invent solutions to problems is central to computational thinking. He argued that computational thinkers can express themselves and their ideas in computational terms. He explained that meaningful expression requires developing both concepts and capacities. He pointed out that capacities for design and social cooperation are often neglected in school. Yet the capacity to design solutions has become more important as the world has increased in complexity. Students need the capacity to design solutions to personal problems such as determining energy-efficient home heating solutions. Students also have to be able to communicate their designs to others and to benefit from the expertise of multiple participants. As a result, students need a way to design solutions that are rich enough to cope with complexity and interactivity in a manner often associated with computational expression. And the design environment needs to support social cooperation in constructing meaningful expressions. Advances of these kinds are synergistic—computing technology itself opens up new possibilities for widespread cooperation.

---

[1] The Molecular Workbench is available at "Molecular Workbench," website, Concord Consortium, http://mw.concord.org/modeler/index.html. Last accessed February 7, 2011.

Supporting Resnick's emphasis on social cooperation, Jill Denner—a developmental psychologist with Education, Training, Research (ETR) Associates—noted that students program differently in pairs than by themselves. She found that students in pairs spent more time doing programming and housekeeping tasks (e.g., saving and testing their code) than did individuals working alone. She observed that most students find programming in pairs highly motivating. When they collaborate students need to develop the ability to communicate concepts. Similarly, Idit Caperton, founder of the World Wide Workshop,[2] described the supports in the Globaloria approach for collaboration and community. Globaloria participants develop original games and publish them on a community Wiki. Participants in the Globaloria community—teachers, students, staff, and teams—all maintain public blogs as design journals, share resources, and build on the products of their peers.

Danny Edelson, director of the National Geographic Society's Geo-Literacy Program, argued that systems thinking is an essential activity of computational thinking. Edelson drew insight from his work in promoting geo-literacy. He noted that geo-literacy calls for a systems view of the world—an understanding of the world as a set of interconnected human social systems and physical environmental systems—and that computational thinking about complex problems calls for a similar understanding.

Jim Slotta, a professor at the University of Toronto's Ontario Institute for Studies in Education, echoed the point that understanding complex systems requires computational thinking. He mentioned a Web-based Inquiry Science Environment (WISE)[3] unit that uses scientific visualizations of global climate change to engage students in reasoning about how their own activities affect the accumulation of carbon dioxide. He noted that the visualizations were designed by Robert Tinker using NetLogo, a language created by Uri Wilensky.

2.3 CONTEXTS FOR COMPUTATIONAL THINKING

Most workshop participants echoed the notion articulated in the first workshop that the power of computational thinking is best realized in conjunction with some domain-specific content. Thus, to understand the human genome, individuals need to combine computational thinking and concepts in genetics. The diversity of possible contexts in which computational thinking applies illustrates its power. Computational thinking

---

[2] Globaloria is available at "Globaloria," website, World Wide Workshop, http://www.worldwideworkshop.org/programs/globaloria. Last accessed February 7, 2011.

[3] "Web-based Inquiry Science Environment," website, University of California, Berkeley, http://WISE.berkeley.edu. Last accessed February 7, 2011.

occurs in a vast array of domains. Developing expertise in computational thinking involves learning to recognize its application and use across domains.

2.3.1 Everyday Life

Many participants provided examples of the use of computational thinking in everyday situations. Troubleshooting devices such as computers, cell phones, and digital cameras involves knowing how to return to a known state (typically by turning the device off and restarting) or test boundary conditions (such as interactions between two applications).

Joyce Malyn-Smith, strategic director of the Workforce and Human Development Program for the Education Development Center, Inc., noted that today's youth carry their technological learning environment continuously in the form of cell phones, computers, and gaming devices. Schools are finding ways to engage students in using their devices to advance learning such as by having them take digital photos for science projects. After school, students bring their devices to community-based programs where they can engage in science inquiry and to museums where they play with exhibits. Taylor Martin, an associate professor of education at the University of Texas at Austin, supported this point, arguing that schools and after-school programs can exploit computational tools such as Facebook.

Lou Gross, at the University of Tennessee and also director of the National Institute for Mathematical and Biological Synthesis, emphasized the value of incorporating a computational worldview into the everyday experiences of students. To illustrate, Gross described an everyday problem—how to pick a checkout line at a grocery store. Gross asked participants to generate parameters that might affect one's decision. Workshop participants suggested line length, the presence or absence of a bagger, someone writing a check, the number of items in a person's cart, and whether the line is an express line. Gross pointed out that high school students often include the presence or absence of someone cute in the checkout line, thus illustrating the point that the criteria for decision making depend on the nature of the model involved and its purpose.

2.3.2 Games and Gaming

A number of participants described game playing and game development as activities well suited to developing computational thinking. They stressed the importance of games that involve domain-specific ideas such as simulations of political situations.

Jill Denner argued that the programming of computer games connects to computational thinking in several ways. Computer games provide a context for the modeling of abstractions. For example, students might program a model of their make-believe world, create variables and new methods, and think at multiple levels of abstraction. They might consider how a player will interact with the game or conceptualize the goal of the game.

Idit Caperton argued that it is possible to learn any subject and to master complex topics or social issues by creating functional, representational, educational multimodal computer games in that domain. The Globaloria environment supports the collaborative development of games and also provides an opportunity for students to play each others' games. For example, Globaloria provides a unit on game design, in which students design an original game about a complex topic (science, math, health, civics) and a social issue that matters to them. Students come up with an idea, assemble teams, and do research. Another Globaloria unit focuses on game development: students develop original game concepts, create prototypes, and produce a complete, playable interactive game. Each unit contains a structured set of learning topics, and each topic contains projects and assignments for students to complete. Assignments scaffold[4] students to create critical parts of their own games.

2.3.3 Science

Robert Tinker advocated the use of simple models of physical phenomena such as temperature, light, and force to teach computational thinking. He described activities in which students use temperature probes to capture data and use graphing programs to develop a model to explain their data. A student makes progress by manipulating and refining the model to reflect increasingly sophisticated understanding of the scientific concepts. The student learns both about the physics of the phenomena and about computational thinking.

John Jungck, at Beloit College and also founder of the BioQUEST Cur-

---

[4] According to Susanne P. Lajoie, 2005, "Extending the Scaffolding Metaphor," *Instructional Science* 33:541-557 (https://www.tlu.ee/~kpata/haridustehnoloogiaTLU/scaffoldinglajoie.pdf; last accessed May 20, 2011). "The term 'scaffolding' was used by Jerome Bruner (Wood et al., 1976) to describe the process in which a child or novice could be assisted to achieve a task that they may not be able to achieve if unassisted, until they are able to perform the task on their own. This definition was influenced by Vygotsky's (1978) conception of the zone of proximal development which is 'the distance between the actual developmental level as determined by independent problem solving and the level of potential development as determined through problem solving under adult guidance or in collaboration with more capable peers' (p. 86). The implication is that individuals have learning potential that can be reached with scaffolding provided by tutors, parents, teachers, and peers."

riculum Consortium, explained the interconnections between computational thinking and biology and described activities that engage students in linking the two. He pointed out that modern biology is essentially an information science. Today, biological data—environmental data and genomic data, for example—is multivariate, multidimensional, and multicausal, and it exists at multiple scales in enormous volume (increasing at terabytes of data per day). He noted that in BioQUEST,[5] students pose original problems, iteratively apply computational thinking to solve those problems, and persuade their peers that their solution is useful and valid. For example, in one activity students pose problems about evolutionary similarities among genes. Using powerful databases they can align multiple sequences of the same gene from different organisms onto one three-dimensional structure. They iteratively refine their representation to illustrate evolutionary conservation across organisms. They use their representation to clarify the comparative biology of sequences in terms of structure, function, and phylogeny.

Walter Allan and Jeri Erickson described computational thinking in ecology and environmental science using a modeling approach. Using simulations to address topics found in the curriculum, they created activities to help students understand complex systems. For example, their Runaway Runoff simulation called for students to conduct experiments on phosphorus pollution using a simulated lake ecosystem.[6] This simulation depicts a lake ecosystem, with fish, zooplankton, and algae that are visible to students as well as bacteria that are invisible to students. Students conduct experiments to develop a food web for the ecosystem. They examine the contents of the digestive tracts of the trout and zooplankton to see how changes in phosphorus affect the populations in the lake and the concentration of dissolved oxygen. They predict the impact of increasing levels of phosphorus on the different populations of fish and zooplankton.

Allan and Erickson explained that the activity scaffolds students to follow a cognitive pattern. This pattern features the same iterative refinement approach described by Jungck. Students start by making a prediction about how a system works. They use a simulation for testing, tinkering, and playing. They record their observations, refine their model of how the system works, and make further predictions. They summarize their findings in essays or posters that describe how runoff affects lake

---

[5] BioQUEST Curriculum Consortium is available at "BioQUEST," website, BioQUEST Curriculum Consortium, http://bioquest.org/. Last accessed February 7, 2011.

[6] A sample student worksheet from the project can be seen at "Runaway Runoff Exercise 1: Who's Who," Worksheet, available at http://simbio.com/files/EBME_WSExamples/RunawayRunoff_WkSh1_example.pdf. Last accessed February 7, 2011.

ecology. These artifacts show that the students learn to make fairly sophisticated models of the lake ecosystem.

Allan and Erickson also described how they implemented this pattern in the "Program a Bunny" environment. In this environment, the bunny is an agent that the student programs to find and eat carrots in a field. The environment is probabilistic, so that carrots are not always located in the same places in the field. A program for a successful bunny must account for the randomness in the bunny's environment. Students can test different programming strategies in a number of increasingly complex scenarios and refine their program. The initial "out of the box" solution is, by design, inadequate for bunny success. Thus, students must learn to modify the program. Modification of the program initiates a cognitive cycle similar to that of the Runaway Runoff simulation involving iterative refinement. The student observes the bunny's success in finding carrots, develops a model of how the program works, and then thinks of another modification that is intended to further improve the bunny's performance.

Lou Gross illustrated ways to use environmental science as a platform for computational thinking. Beginning with an aerial image of Washington, D.C., from Google Earth, students are asked, How would you describe this image? After listing typical topographic features such as buildings, roads, and trees, students eventually describe the image by saying how much of the image is this color or that color, how much is made up of buildings, how much of roads, and so on. Gross described these observations as the basis for describing the image as a vector where the components consist of the fraction of the image that is of each type. One interpretation of this vector is that it represents a probability distribution of the landscape for a discrete number of components. Students are scaffolded to realize that spatial aspects of the image are not included in the vector description. They explore how some large-scale temporal variations (such as the growth of cities) could be captured by a time-varying vector. This activity prepares students to use prepackaged software to take advantage of computational methods for looking at change across a landscape, e.g., coupling between an image, a dynamically changing vector, in this case a bar graph, and then an overall descriptor.

Danny Edelson showed how geography and earth science involve computational thinking. Edelson described some of the issues that arise for students learning to understand geographic data:

- *Continuous versus discrete data sets.* Students learn about the issues in transforming a map from a continuous representation to a map represented in discrete pixels or cells. They can articulate all the positive and negative implications of each representation and learn how the representation affects the results of their data analysis.

- *Color representations of temperature.* Students explore the implications of using color to represent temperature on a map. For example, although it makes physical sense to subtract two temperatures (e.g., January's temperature from July's temperature), it does not make much sense to subtract yellow from red. Color representations on the map cannot be manipulated in the same way as the underlying physical parameters. To resolve the paradox, students need to realize that temperature maps consist of regular arrays of numerical data. They can be transformed into colors, but their underlying mathematical character is maintained.
- *Boolean operations.* Boolean operations are key analytic tools for interpreting maps. Students gain insight into Boolean operations by testing and refining solutions to problems. For example, to analyze geographic data students might be asked to find counties in the United States whose African American population exceeds the Caucasian population.
- *Spatial relationships as specifications of sets.* In working with geographic data, a student might want to find the intersection of two regions on a map, where the regions are specified according to some nonspatial criteria. Managing such operations intellectually calls for thinking about them as combinations in one sense and as spatial entities in another sense.
- *Satisfaction of multiple constraints in problem solving.* Students might be asked to locate a power plant in areas that are both accessible to railroad transportation and close to large bodies of water. Students learn how to use logic tools to locate specific geographic features.

Robert Panoff advocated teaching computational thinking through computational science, in part because this approach develops metacognitive skills or the ability to monitor understanding of computational results. Panoff drew on quantitative reasoning and multiscale modeling to illustrate various anomalies in how people conceptualize quantity. Examples include:

- *Inconsistent and faulty intuitions about numbers.* Many people believe that two-fifths (2/5) is a small number, whereas 40 percent feels like a large number to them. Panoff noted that one metropolitan police department assigned more officers to patrols on Friday and Saturday night because a careful analysis of the data showed that just under 30 percent of car break-ins were on either a Friday or a Saturday night. Since 2/7 is 29 percent, the frequency of car break-ins was actually consistent across weekdays and weekends!
- *Representations of numbers in computational media.* In principle, the arithmetic expression given by $355/113 - 101/113 - 101/113 - 101/113 - 52/113$ should equal zero. But when the expression is evaluated on most calculators, a non-zero result is obtained. Because most students realize

that "something's not right" when they are confronted with this expression, such a realization can be the beginning of a serious exploration of how numbers are represented in a computer.

- *Interpretation of orders of magnitude.* Panoff noted that many people have difficulty recognizing the degree of precision necessary to make an inference. He illustrated the point by asking what a student needs to know in order to answer the question, How much bigger is Earth than Pluto? An obvious way to approach this problem is to perform Internet searches for the mass of Earth and the mass of Pluto. But an Internet search for the mass of Earth generates 20 or 30 different values, which have a spread of several percent. How does one know which value to use or how to proceed? Here context matters—why is one asking the question about relative sizes? If the question relates to how big an object has to be in order to be a planet, then in the absence of a formal definition of "planet," one needs to know only that the ratio M_{Earth}/M_{Pluto} is on the order of a few hundred—and a difference of "several percent" is simply irrelevant to knowing which value of M_{Earth} to use.

2.3.4 Engineering

Christine Cunningham, vice president at the Museum of Science, Boston, described engineering as a focus of computational thinking for elementary education. Echoing discussions from the first computational thinking workshop, she pointed to intellectual parallels between computational thinking and solving engineering problems. Cunningham stressed that understanding engineering habits of mind and mental processes is an important goal of elementary science. She illustrated how these habits of mind require important aspects of computational thinking. Cunningham identified 20 topics that are commonly covered in elementary science programs, paired each with an engineering specialty, and illustrated the pairing with a particular technological device or process. For example, environmental engineering can be introduced using water filtration devices to help students understand the human impacts on the water cycle. In another example, a solar cooker can illustrate principles of energy and connect to sustainable engineering.

2.3.5 Journalism

Ursula Wolz described the use of the language arts and journalism as a vehicle for exploring computational thinking. She argued that insights into computational thinking can come from comparing the precision of computer languages to the challenges of precise communication in journalism using natural language. Journalism involves principled storytelling and information dissemination. Journalism students are constructors of aggregated content (rather than just consumers). To produce a story, students must inquire, create, build, invent, iterate on the account, polish, and publish. Wolz emphasized that students iterate on defining the problem, researching it, drafting a solution, and testing it. They copy edit and fact check. In the end, they publish and get more feedback. All of these same notions arise in other instances of computational thinking.

2.3.6 Abstracting Problem Solving from Specific Contexts

Given the diversity of contexts discussed and even the diversity of problems within a single context, a number of workshop participants discussed the use of computational thinking across contexts or topics. For instance, several noted that individuals were likely to need different (though overlapping) sets of computational thinking skills. Thus, physicians need to learn how to use visualization tools, as do teachers. Joyce Mayln-Smith of the Education Development Center suggested that the computational thinking abilities needed by users of information technology tools and applications are different from those needed by producers or developers of such tools and applications. Consequently, the pedagogical approaches needed for developing these skills must be suited to the goal.

Michelle Williams, assistant professor of science education at Michigan State University, made a similar point, arguing for helping students and their teachers recognize that the computational thinking skills they use to make sense of representations of scientific knowledge work for multiple representations. Williams showed how a WISE project can scaffold students to use computational thinking skills as they engage with a number of computer-based representations. In her project students used simulations of mitosis to understand phases of cell division, and Punnett squares to determine the genotypes and phenotypes of different generations of plants, and they interacted with the Audrey's Garden animation[7] to make distinctions between inherited and acquired traits.

---

[7] See "Case of Audrey," website, Exploring Younger Children's Understanding of Heredity, http://education.msu.edu/research/projects/nsf-heredity/curriculum.html. Last accessed March 14, 2011.

2.4 PEDAGOGICAL ENVIRONMENTS FOR COMPUTATIONAL THINKING

2.4.1 Foci for Pedagogical Environments

Many presenters stressed that interactive visualizations or simulations are at the heart of computational thinking. This perspective cut across multiple disciplines. For example, John Jungck argued that the key to computational thinking in a biological context is in the power of visualization. Robert Tinker stressed the value of visualizations in simple models of scientific concepts such as temperature, light, and force to teach computational thinking. (See Figure 2.1.) Danny Edelson argued that Earth models are best understood in terms of dynamic and spatial models, and he illustrated the point using a NetLogo model for infiltration and runoff processes in a region in the presence of precipitation. Jim Slotta pointed to visualizations of global climate change. Michelle Williams emphasized visualizations in understanding genetic inheritance. Mike Clancy, senior lecturer in the Department of Computer Science at the University of California, Berkeley, suggested that the causal relationships depicted in models are similar to the causal relationships entailed in understanding what a computer program actually does in execution, so that, for example, a student needs to understand what causes a program bug or a program to perform in a certain way.

A second focus for pedagogical environments is the modeling and troubleshooting of data sets. For example, Robert Panoff noted the importance of understanding limitations in the underlying data. Danny Edelson noted that anomalous data often catch people's attention and generate a teachable moment by motivating them to understand the cause of the anomaly. Panoff stated that even experts sometimes miss anomalies. Uri Wilensky of Northwestern University stressed the advantages seen when students collect data themselves and then use those data to try to fit models to those data, rather than using data provided by others. Christina Schwarz of Michigan State University illustrated how students benefit from iteratively refining their models of data they collect.

A third focus is searching for patterns in large data sets. John Jungck illustrated the forms of computational thinking students use to explore, analyze, interpret, and synthesize massive amounts of biological data. Similarly, Stephen Uzzo, vice president of technology at the New York Hall of Science, pointed out that e-science, which focuses on managing, modeling, and making discoveries in massive amounts of captured data, seeks patterns, dynamics, influences, and complex and emergent behav-

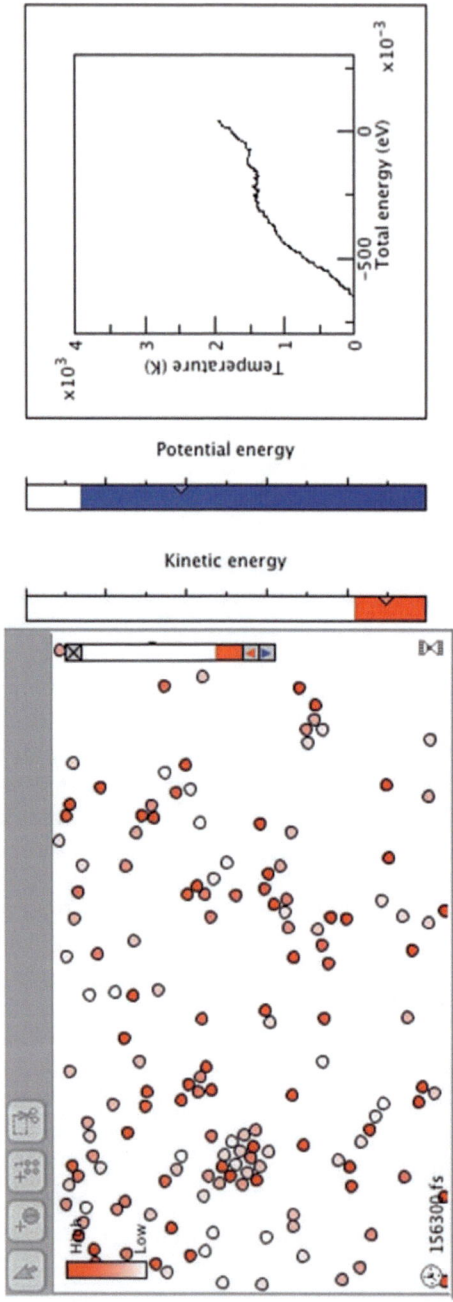

FIGURE 2.1 Molecular Workbench example—one step in integrating computational thinking for using computational thinking in science. A learning progression in science might start with learning that numbers are associated with physical properties like temperature, and end by having students create computational models of these properties. Along the way an important step is using computational models and seeing that these models are influenced by parameters. This figure is a snapshot from a Molecular Workbench (http://mw.concord.org/modeler) molecular dynamics model of phase change showing a liquid-gas mixture on the left that has been heated by adding kinetic energy until it is almost all gas. On the right, a graph of average per-atom kinetic energy against total energy shows a distinct plateau where all heating ended up increasing the potential energy, which resulted in pulling the atoms apart. Students can explore the effect on the liquid-gas transition temperature by changing the strength of attraction between atoms by using molecules and changing their number.

ior in whole systems.[8] The computational thinking needed to engage in e-science addresses, among other things, complexity; data visualization; network science (that often results in theoretical generalizations); data interoperability, data sharing, and other collaboration skills; and the use of semantics for searching or creating more effective data structures.

2.4.2 Illustrative Pedagogical Environments

Workshop participants illustrated the potential of pedagogical environments for scaffolding learning, supporting iterative refinement, and enabling students to use visualizations or large data sets. Several alternative approaches were described.

- *WISE (the Web-based Inquiry Science Environment)*. Jim Slotta described how WISE is designed to provide scaffolding for inquiry activities in science classrooms. WISE guides students to engage in computational thinking using scientific visualizations, simulations, models, complex data, or long-term projects. A typical WISE project might engage students in designing solutions to problems that require computational thinking (e.g., design a desert house that stays warm at night and cool during the day using simulations of the day/night cycle and other resources), debating contemporary science controversies using a variety of evidence (e.g., the causes of declining amphibian populations using systems thinking), critiquing scientific claims based on modeling (e.g., arguments for the occurrence of global climate change), or conducting virtual experiments (e.g., exploring a simulation of airbag deployment and studying factors such as velocity and driver height). Tools and interactive materials provided in the WISE environment support collaborative activities and cognitive guidance to promote reflection and critique.

- *Scratch*. Mitch Resnick described a computational environment called Scratch for facilitating individual expression. The MIT Media Lab developed Scratch as a companion to an online computational thinking community to help engage people in creative learning experiences and to facilitate creative thinking. A graphical programming language, Scratch gives its user the ability to build programs by snapping graphical blocks together. The blocks control the actions of different dynamic actors on a screen. This approach to program construction enables users to avoid issues of syntax and other details that often distract them from the critical processes of designing, creating, and inventing. Using Scratch, it is easy to engage in iterative refinement. Scratch also facilitates social cooperation

---

[8] See, for example, Defining e-science, website, U.K. National e-Science Centre, http://www.nesc.ac.uk/nesc/define.html. Last accessed February 7, 2011.

FIGURE 2.2 Scratch programming example—a screen-shot of a game made in Scratch. The goal of the game is to maneuver the big fish to eat the small fish. The programming scripts for the big fish are in the middle panel of the screen-shot. This program highlights several concepts, including the idea of an algorithm (e.g., iteratively point toward the mouse-pointer and take a few steps toward it); computational concepts like variables (shown in the "score" variable); and computational practices like modularization and abstraction (shown in the scripts for "increase-score" and "mouth-animation"). The graphical nature of the Scratch programming language makes it easier to focus on computational concepts and practices without worrying about syntax issues (which have no conceptual value). Also, the nature of the Scratch interface makes it easy to build up scripts in an incremental, iterative style aligned with modern software development practices.

by making it very easy for a user to share his or her design with others for comment and feedback. (See Figure 2.2.)

  • *Storytelling Alice.* Storytelling Alice is a programming environment designed to motivate a broad spectrum of middle school students to learn to program computers through creating short three-dimensional animated movies.[9] Jill Denner engages students with a use-modify-create approach. First, students play other students' Alice games and work through three

---

[9] Caitlin Kelleher, 2007, "Storytelling Alice," website, Alice.org, http://www.alice.org/kelleher/storytelling/. Last accessed February 7, 2011.

tutorials that illustrate how to program with Alice. The goal of the "use" phase is for students to learn about the Storytelling Alice interface and the kinds of games that they might make. Second, students learn to modify an existing game through a series of graduated self-paced challenges. The goal of the "modify" phase is to understand the mechanisms that they will use to program a game. Third, a student creates an original game using the language. Students work with specific computational thinking concepts such as event handling, parallelism, additional methods, parameters, alternation, iteration, and conditional execution, and many of Denner's students created their own methods and used parameters that illustrate the use of modeling and abstraction.

- *Globaloria*. Idit Caperton described Globaloria as a project-based learning environment for stimulating computational thinking, creativity, and inventiveness in youth and educators. Globaloria supports project-based, multidisciplinary, innovative and creative learning (of any subject) through software design and programming. Globaloria emphasizes six abilities, including the ability to program an educational game, wiki, or simulation; to use project management skills in developing programmable wiki systems in a Web 2.0 environment; to produce animated media and to program, publish, and distribute interactive purposeful digital media in social networks; to learn in a social constructionist manner and to participate actively in the exchange of ideas and artifacts; to undertake information-based learning, search, and exploration as they relate to the abilities above; and to surf websites and use web applications thoughtfully as they relate to the earlier abilities enumerated. (See Figure 2.3.)

Many presenters testified to the pedagogical value of features found in the environments described above. Panoff, Tinker, and Jungck, among others, stressed the importance of supporting iterative refinement. Denner, Caperton, Gross, Resnick, and others described sequences of activities that culminated in complex projects.

2.5 DEVELOPMENTAL CONSIDERATIONS AND COMPUTATIONAL THINKING

An important issue for many workshop participants was the development in novices of facility with computational thinking. Presenters offered varied perspectives on how computational thinking might develop and on what constraints should be considered. Those favoring project work generally argued that students working collaboratively on a compelling problem could use much more advanced forms of reasoning than teachers might expect. Others argued that young children have some impressive capabilities but also some very serious limitations.

FIGURE 2.3 Globaloria example—a depiction of the Globaloria learning environment, where students develop computational thinking skills through team design and creation of computer games. Students are tasked with using Web 2.0 tools and media in developing various games on topics such as health, civics, and science.

Robert Tinker noted that because students begin as concrete thinkers, it remains a challenge to identify the age or grade level at which children can handle abstraction. As an example, he said that second graders can hook up a probe to measure temperature and understand how the device works. Fourth graders can demonstrate reliable interpretations of heating and cooling curves and use the results to gain understanding of abstract principles. Even adults, however, often have difficulty with concepts such as thermal equilibrium or insulation. He reported on studies showing that even well-educated scientists were not sure about the value of wrapping a cold drink in a sweater to keep it cold.

The workshop participants argued that many theories of learning are relevant. Participants mentioned theories of multiple intelligences, knowledge integration, experience-based learning, how novices and experts learn, and how groups learn.

2.5.1 Development of Scientific Thinking in Children

Deanna Kuhn, a developmental psychologist at Columbia University, spoke about the evolution of young learners through different intellectual stages with respect to scientific thinking. She focused on how they use data and evidence and on their facility with scientific thinking.

For Kuhn, the first accomplishment of development involves accepting the possibility of false belief. The child must conceive of data as possibly not representing the complete reality. Very young students can recognize simple covariation in causal models; that is, Did A cause O? in simple cases. But they have difficulty with covariation in a multivariate context (e.g., Which A caused O?) or a negative antecedent and outcome (e.g., Not A and O). Such students often interpret data in isolation and do not look for large patterns.

Kuhn pointed out that although young children can do fundamental experimental design, they often close their inquiry prematurely. Premature closure sometimes occurs when children are presented with confirming evidence. Children often stop the inquiry at this point, not realizing that the inquiry remains unfinished and that confirming evidence is not sufficient to rule out competing hypotheses. Mitch Resnick noted that even adults identify one cause (of potentially many) and assume that a partial inquiry is completely explicative.

Although young children can successfully employ some of the intellectual skills of scientific thinking, they can have a hard time articulating how they know something. In particular, they do not understand the epistemological difference between claim and evidence. For example, looking at a photo of a boy standing on an award podium with a sign labeled with the number "1" and holding a trophy, a child is asked, "How do you know that this boy won the race?" A child will often answer not with evidence of how she knows (e.g., "He is holding a trophy or the podium has a number 1 on it") but with a theory of why the outcome makes sense (e.g., "His sneakers were fast").

Students as old as 12 sometimes focus on evidence and data fragments that support their story, while ignoring or minimizing those that do not. For example, in explaining what causes an avalanche, a student may report that in case A, it was the slope angle that caused an avalanche. Yet the same student will claim that in case B, the slope angle did not make a difference because the slope angle was small and something else caused the avalanche. These students are having trouble distinguishing between a variable and a variable's magnitude. The educational challenge at this level is to help the child see the data as evidence rather than as an example of a favored claim. Kuhn argues that when a child develops a sort of meta-awareness (control) over this sorting and attribution process, true scientific thinking can begin. Several participants noted that even graduate students sometimes ignore contradictory evidence and focus on supportive evidence.

Jill Denner reported on several lessons learned from her research. For example, she pointed out that individual differences matter a great deal, because individual students have different starting levels, willingness to fail, and motivations. Students are sometimes more comfortable, sometimes less, with the idea of following step-by-step instructions to carry out a task. Some students are more afraid than others to fail and thus are unwilling to tackle problems that entail the risk of failure (e.g., of using a concept incorrectly). Denner found it necessary to balance student engagement on a problem with motivating students to learn more complex or difficult concepts needed for their programs.

Denner also pointed out that students with poor reading skills face special challenges. Although the challenges of modifying programs are a good way to ease into game programming, understanding an existing program is a text-heavy exercise and thus is difficult for English-language learners and students who have reading difficulties.

2.5.2 Possible Progressions

As a preliminary point, a number of workshop participants felt that it is often possible to get students to use even advanced computational thinking without invoking the use of that term. For example, Robert Panoff argued that once students are thinking about a leaky bucket as a time and rate problem, they are in fact doing calculus. When they take the difference between two things and divide by the interval between those two things, they are taking derivatives. When they're averaging, they are doing integration. Panoff's philosophy here is to help students break a big problem into smaller problems and let the computer do the small parts—a process that helps to empower students. Taylor Martin argued that educators could see this approach as being a "sneaky" way to get students to use computational reasoning. Once engaged, students continue to use computational thinking and even begin to see its applications across contexts.

As a point of departure for considering learning progressions, Joyce Malyn-Smith proposed a sequence:

- *Grades K-4*, to focus on computational thinking literacy, career awareness, and computational thinking skills for learning. An overarching theme in this time frame might be the lesson that learning is cumulative—a student can learn more by building on something he or she already knows.
- *Grades 5-8*, to continue to address computational thinking literacy but add career exploration and learning about computational thinking skills for various careers in science, technology, engineering, and mathematics (STEM). This exploratory phase would offer students an opportunity to test their interest in various careers.
- *High school*, a final preparatory phase, to prepare students to have the credentials to be able to keep doors open so that they can move into computing careers and careers in other STEM fields in which computational thinking will give them really strong opportunities.

Others, including Panoff, Allan and Erickson, and Denner, proposed looking at a learning progression for the development of computational thinking by applying the use-modify-create continuum over and over again. For example, a student first runs a model to see what happens. Then she may modify it by moving a slider bar, or two or three slider bars. And then she may change the number of slider bars. Finally, she writes a model that calls for the use of slider bars to change parameters. By iterating on this pattern, the student gains progressively more capabilities in the area of computational thinking.

Peter Henderson would start with computational thinking activities involving pattern recognition and naming in pre-K, although for the first several years, the term would not be introduced explicitly. Only later would the notion of computational thinking be explored as such. In this sequence, traditional mathematics, discrete mathematics, and logical reasoning are taught at all grade levels. This has important implications for high school, where an advanced placement (AP) course in discrete mathematics would replace the current AP course in computer science. A freshman discrete mathematics sequence would be introduced, similar to that currently present for calculus. This approach would allow students to link mathematics and science following the traditional engineering educational model. This model emphasizes the connections across the science and math foundations of the disciplines (e.g., physics, chemistry, calculus). Clancy pointed out that this approach could also apply in college course sequences.

2.6 ASSESSMENTS FOR COMPUTATIONAL THINKING

Many workshop participants stressed the importance of student evaluation for pedagogical purposes. For example, Christine Cunningham pointed out that both teachers and students in the Museum of Science's Engineering is Elementary project pay much more attention to material when student understanding of such material will be evaluated. She cautioned that narrow goals for evaluation are counterproductive. Students and teachers need to appreciate the links among topics, and goals for courses need to acknowledge these dependencies. If students and teachers know that an evaluation will involve student knowledge of, for example, looping, they proceed to learn and teach looping. However, if teachers and students realize that an evaluation will involve student knowledge of program design, and knowledge of looping helps students understand program design, both students and teachers are more likely to connect looping and program design.

To evaluate the connections and interdependencies in computational thinking in introductory programming courses, Mike Clancy uses a case study approach and lab-centric instruction. A case study is a worked-out solution accompanied by a narrative of how the solution was identified. The narrative discusses design tradeoffs, evidence for alternatives, methods for testing the solution, debugging, and other issues such as optimizing. Students respond to questions that require them to consider new alternatives, critique design choices, develop test suites, and interpret results of tests conducted by others.

Lab-centric instruction emphasizes hands-on lab hours supervised by a teaching assistant rather than lecture and discussion. But because there is more lab time than in most lecture/discussion courses, the course has room for a number of embedded assessment activities. Lab instructors can also monitor most of what the students are doing, and have a window into much of their thinking and not just their finished work. Thus lab instructors can notice confusion when it occurs and address it immediately to provide targeted tutoring. Clancy reported that students in lab-centric courses are less likely to drop the course, possibly because their confusions are caught before they become too burdensome.

2.7 TEACHERS AND COMPUTATIONAL THINKING

Teaching computational thinking requires both knowledge of the discipline and skill in teaching when students collaborate to solve complex problems (sometimes referred to as pedagogical content knowledge). John Jungck argued that the primary challenge for teachers of computational thinking is placing student interests at the center of problem posing. In courses where students pose and solve problems teachers lose much of the control they traditionally have over the learning process and may become uncomfortable. They need new skills to guide individual learners. Supporting students engaging in self-directed collaborative processes requires an ability to diagnose difficulties and give hints rather than supplying solutions. Designing effective assignments is also challenging, but many programs such as BioQUEST offer excellent options.

Michelle Williams stressed that to be effective, teachers have to understand where students are starting. Further, teachers need to determine the types of understandings that students must have to be successful and to design new ideas or computational activities to provoke students to engage in computational thinking.

Jungck noted that students in some cases may have more technical skills than their teachers in the area of using computers. Williams pointed out that teachers often find ways to make individual students class "experts" on troubleshooting the operating system or accessing online materials to take advantage of available technical skills. Williams also noted that teachers need professional development to become proficient in teaching computational thinking. In her work she found that teachers followed a learning progression, becoming more proficient over time in using technology and guiding students with inquiry questions. Thus teachers of computational thinking may well be called on to assume new and unfamiliar roles in the classroom and need support to become proficient in performing these roles.

Cunningham argued for the importance of building on what teachers know or feel comfortable doing. It is well known that many elementary school teachers are uncomfortable with science because of their limited preparation in this area. Cunningham argued that engineering (and presumably computational thinking) is even more terrifying. To build on what teachers know, Cunningham and colleagues begin their professional development by connecting exercises in literacy—an illustrated storybook for children—with engineering. The story has significant engineering content, but it is presented as a reading exercise so that teachers can use established skills to lead their classes. Students receive a very general introduction to engineering and to some computational thinking concepts from the book. The book provides context for the hands-on engineering activities that the kids will be doing in their classes.

2.7.1 Professional Development and Other Needs for Teacher Support

Participants described a number of alternative views concerning methods and models for professional development. Cunningham suggested starting small. Teachers tend to be more willing to invest a couple of class periods to experiment with a new concept, rather than an entire school semester or year. The success of one individual teacher with a particular concept or topic can catalyze others, as his or her students tell their friends about an interesting new experience in class. Other teachers also hear about such a program and often want to try it themselves. These efforts build grassroots support for change.

Jim Slotta agreed that teachers are more willing to use materials for a short period to see if their students benefit from a particular approach. He described the experience of the Technology-Enhanced Learning in Science (TELS) center, where teachers first used a 1-week unit featuring visualizations. He also noted that asking teachers to identify the topics for professional development was effective. Initially teachers asked for help with the technology. These issues were resolved, and the teachers then asked for guidance on using inquiry questions. Next teachers asked for help with using visualizations. Successful professional development involved making videos of varied teaching practices and conducting a dialog where teachers discussed the alternatives and identified a set of best practices.

Jill Denner reported a number of challenges in promoting computational thinking in middle school. These included mundane issues such as difficulties with hardware and software and with Internet access, consistent with the comments of Slotta. Taylor Martin emphasized that access to computers and provision of technical support are important, stressing that computers are the tool students will use in the workplace. Teaching computation without them is not really preparing students for the real world. Many schools lack access to computers or only have productivity tools like word processing rather than the computational environments mentioned in the workshop (e.g., WISE, Scratch, or Globaloria).

Several participants emphasized the importance of combining professional development with solid curricular materials. Because precollege teachers are often inexperienced with the subject matter of engineering, teaching materials have to be explicit and clear. Cunningham argued that when learning objectives drive the experiences embedded in curricular units, objectives need to be very explicit and specific rather than high-level and abstract. She argued that learning objectives should also be few in number and relatively narrow so that a high degree of student success is possible. She suggested that the materials provide ways of specifically assessing the scope and extent of student mastery and comprehension.

Cunningham and colleagues have found that hands-on experiences are particularly important for young learners. They have fielded many requests to replace physically manipulative experiences in handling objects with a click-and-drag interface on the computer that students can use to connect objects on the screen. But knowledge about the physical world that teachers take for granted cannot be assumed in students. For example, students don't necessarily know that a fuzzy pompom will pick up pollen better than a smooth marble. In fact, that fact is engineering knowledge, and it's "common sense" only if one has real-world experience with pompoms and marbles.

Experience with the physical world varies across populations. Cunningham reported that many students, including especially girls and underrepresented minorities, lack cultural experiences that illustrate the value of learning about engineering or the benefits of advances in engineering. She and her colleagues use hands-on materials as well as storybooks that bring these ideas to life.

Williams reported on her experience working with precollege teachers. She stressed the importance of engaging teachers in reviewing student work. She has found it valuable to have teachers use the scoring rubrics developed by the curriculum designers. She observed that teachers can make big gains in the sophistication of their teaching by making changes based on the gaps in their students' knowledge.

2.7.2 Teachers and Career Awareness

Joyce Malyn-Smith pointed out that teachers can play an important role in helping students make connections between what they know and what they are learning. Teachers can encourage students to connect the new ideas to activities they would like to perform either in the present or in the future. Teachers can help students understand the connection between computational thinking and future earning power. Malyn-Smith said that students often have understanding of details about computational thinking from their areas of interest but lack the historical and cultural frameworks for placing such information in context. Teachers can help students to validate what they know and to understand how it is important and how it relates to what they are learning in class.

Williams added that instructional materials that connect to personally relevant problems can help teachers make connections between science and students' ideas. Questions such as determining the origin of one's eye color or distinguishing among possible ways to reduce the accumulation of greenhouse gases stimulate exciting conversations between students and teachers.

2.8 LEARNING CONTEXTS AND COMPUTATIONAL THINKING

Workshop participants contrasted formal, informal, and ubiquitous learning contexts. They noted that computational thinking may fit better in some contexts than others.

2.8.1 Aligning with Standards

Several presenters stressed the challenges posed by a tightly packed curriculum that does not necessarily stress abstract thinking skills but that could provide a framework for integrating curricular content. Cunningham underscored the importance of integrating the new material—in this case, engineering—with what schools are already teaching. Successful integration can show how the new material contributes to understanding. Arguing for new material as a primary focus, however, is not likely to succeed because of preexisting curriculum demands.

Cunningham noted the importance of articulating how new content and skills in the Engineering is Elementary project connect to existing educational standards. Such connections could include, for example, core concepts of technology such as systems, processes, feedback, controls, and optimization; the design process as a purposeful method of planning practical solutions to problems; inclusion in the design process of such factors as the desired elements and features of a product or system or the limits that are placed on the design; and the need for troubleshooting.

Paulo Blikstein of Stanford University noted that often typical instruction is oriented toward declarative knowledge, whereas computational thinking is oriented toward procedural knowledge. In this view, declarative knowledge provides content (and is essential to particular fields or careers), whereas computational thinking is most useful for integrating and building connections in the midst of such knowledge. Those accustomed to thinking primarily in terms of declarative knowledge may find it difficult to appreciate educational themes oriented toward procedural knowledge.

Allan and Erikson reported that the development effort for the EcoScienceWorks[10] project approached the use of programming instrumentally. Downplaying the use of programming was a response to the developers' concern that some teachers might rebel because the Maine learning standards did not mention programming. They feared that they would have a hard time justifying spending scarce classroom time on program-

---

[10] EcoScienceWorks is available at "EcoScienceWorks: Exploring and Modeling Ecosystems Using Information Technology (IT)," website, Foundation for Blood Research, http://www.fbr.org/swksweb/esw.html. Last accessed March 14, 2011.

ming, even if focusing explicitly on programming might have significant educational value.

Robert Tinker observed that students involved in a very tightly packed K-12 curriculum do not have the time to master programming. His preferred approach is thus to use a programming environment such as NetLogo[11] or AgentSheets[12] that allows users to focus on the concepts represented rather than on the details of programming. Janet Kolodner of the Georgia Institute of Technology noted that another option is the use of powerful software suites in which the student can manipulate important parameters.

Several participants noted that learning about engineering or computational thinking may meet teacher goals that are not necessarily based in educational standards but are expected outcomes for students. For example, Cunningham observed that many elementary school teachers want to find ways to help their students work together in teams. Persuading students to work together, to respect each other, and to communicate what they're doing is something that many teachers want to accomplish at the beginning of each year, because learning to work in groups is a skill elementary teachers are expected to develop in their young pupils. Educational activities that call for collaboration can often be an important part of such persuasion.

2.8.2 Out-of-School Computational Thinking

Given the issues relating to introducing computational thinking into schools, a number of workshop participants pointed to out-of-school venues as providing significant opportunities for exposure to computational thinking. In out-of-school venues, students have the time to engage in complex projects that are needed to nurture computational thinking. Malyn-Smith noted that learners need opportunities for thoughtful, reflective engagement with phenomena—not just a "drive-by" experience. Teachers in Malyn-Smith's program are encouraged to think broadly about the knowledge base that students are developing in all of their activities, not just those provided in program settings. Teachers also engage in conversations with students about their interests and what they are learning in other settings, such as in museums, in watching television and listening to the radio, by playing games, and through what they're doing with their friends.

---

[11] NetLogo is available at "NetLogo.com," website, Northwestern University, http://ccl.northwestern.edu/netlogo/. Last accessed March 14, 2011.

[12] AgentSheets is available at "AgentSheets, Inc.," website, AgentSheets, Inc., http://www.agentsheets.com/index.html. Last accessed March 14, 2011.

Out-of-school environments can provide curricular flexibility, appropriate staff capacity, infrastructure access, and access to effective programs, Malyn-Smith explained. This is especially valuable in rural areas. These interrelated challenges have constrained many previous educational innovations, and computational thinking is no different, she argued. For example, although nearly every middle school student learns from the textbook that trees help mitigate pollution, students in an after-school program can have a chance to go further, using modeling tools to map the trees in their school yard and to record relevant data on species, health, growing conditions, and the like.[13] Students can use automated models to calculate the benefits of the trees in terms of pollution removal and runoff mitigation, and they can model alternative growth scenarios as they either "plant" new trees, let the existing trees continue to grow, or remove the trees for expanded parking. Re-running the model leverages the power of automation to quickly adjust the underlying parameters and see what the impacts are. This iterative process just doesn't fit in a curriculum packed with hundreds of discrete topics that are connected loosely at best. Time allocations that allow for depth and complexity are possible in these programs. Schools have to provide this type of time allocation as part of the culture change needed for computational thinking to take root.

Stephen Uzzo promotes computational thinking as a way to help future scientists cope with the transformational effect of data-rich science. New York Hall of Science activities entail developing exhibits, implementing them, and then evaluating them for pedagogical efficacy in conveying the relevant concepts to students.

For example, Uzzo discussed a project developed cooperatively with the School of Library and Information Science (SLIS) at Indiana University. The SLIS macroscope helps to identify patterns, trends, and outliers in very-large-scale static or streaming data sets. The macroscope is an expandable and integrated set of applications that scientists can use to share scientific data sets and algorithms and to assemble them into workflows. Macroscopes continuously evolve as scientists add and upgrade existing plug-ins and remove obsolete ones to arrive at a set that is truly relevant for their work. This project requires little or no help from computer scientists.

Uzzo argued for a new generation of science students who know what it means to be an e-scientist, taking advantage of online data. He suggested that informal learning institutions may be in the best position

---

[13] This example is further elaborated in ITEST Small Group on Computational Thinking, 2010, *Computational Thinking for Youth*, Newton, Mass.: Education Development Center. Available at http://itestlrc.edc.org/resources/computational-thinking-youth-white-paper. Last accessed May 20, 2011.

to advance the cause of e-science. Specifically, he said that informal science institutions have an opportunity to integrate computational thinking in a broad range of science activities. These institutions are in a good position to conduct learning research on computational thinking and to integrate such research into professional development and curriculum development for K-12 formal education.

2.9 RESEARCH AND UNANSWERED QUESTIONS REGARDING COMPUTATIONAL THINKING

The first workshop report identified five open questions that at least some participants in that workshop believed were worth further exploration:

1. What is the structure of computational thinking?
2. How can a computational thinker be recognized?
3. What is the connection between technology and computational thinking?
4. What is the best pedagogy for promoting computational thinking?
5. What is the proper institutional role of the computer science community with respect to computational thinking?

Several of these questions were discussed in the second workshop: question 2 is related to the discussion of student assessment (Section 2.6); question 3 is addressed in Section 2.9.2; and responding to question 4 is implicitly the purpose of Sections 2.3 and 2.4. In addition, participants in the second workshop raised additional issues that are described below.

2.9.1 The Importance of a Process for Defining Computational Thinking

As noted above, the first workshop identified the structure of computational thinking as an important open question. A number of participants in this second workshop amplified this observation by pointing to the importance of a process for defining computational thinking.

For example, Joyce Malyn-Smith argued that the field needs a rigorous and valid way of bringing people together and figuring out what computational thinking is. It is necessary to have consistency because not everyone understands what computational thinking is about, or they see it only through their own lens. Absent a rigorous process for defining computational thinking, efforts to promulgate computational thinking in the curriculum will lack credibility. Whatever else it may be, computational thinking in the curriculum cannot be just a bunch of examples that are placed into the curriculum at the discretion of individual teachers.

2.9.2 The Role of Technology

Elaborating on the first workshop's question regarding the connection between technology and computational thinking, Malyn-Smith identified two research questions. First, to what degree and in what ways does the technology expertise of youth contribute to their computational thinking? A related second question is, How and to what degree can the use of technological tools and systems and processes facilitate transfer of learning in STEM careers and in the sciences?

2.9.3 The Need for Interoperability

Al Aho noted that "the software world of today is largely a Tower of Babel, with lots of incompatible infrastructures and a lot of expense regarding who pays, who collects the data, who maintains the data, who maintains and evolves the software." Stephen Uzzo said this was especially true in an e-science environment in which data is produced in prodigious quantities and there is a premium on making large data sets available to researchers reliably and promptly. In this view, computational thinking efforts would be facilitated by interoperability between applications used by researchers, and it must provide easy-to-use tools for processing, manipulating, and combining multiple data types.

Jim Slotta echoed these points when he observed that content from most platforms is not portable across platforms. Further, the environment of a given platform is generally unable to interact with other applications that are running on the machine. To address some of these limitations, Slotta and his team engaged with the computer science department to develop a new open-source architecture called SAIL (Scalable Architecture for Interactive Learning) for content display and manipulation that separates the various layers of the learning environment (and in particular the content and the user interface) wherever possible.

SAIL has been used in a number of other science education efforts as well. For example, SAIL is an integral element of the Science Created by You (SCY) project of the European Union.[14] SCY is a large project that provides a flexible, open-ended learning environment for adolescents. Within this environment—called SCY Lab—students engage in personally

---

[14] This discussion of SCY includes material found at "Science Created by You," website, http://www.scy-net.eu/. Last accessed February 7, 2011.

meaningful learning activities that can be completed through constructive and productive learning. Examples of such learning activities include browsing for information, generating a hypothesis, and distributing tasks.

Slotta has also developed a technology framework called SAIL SmartSpace (S3) to support a complex orchestration of people, materials, resources, groups, conditions, and so on. This framework can be regarded as a "smart classroom" infrastructure that facilitates cooperative learning in a milieu of physical and semantic spaces. From a technical standpoint, S3 supports aggregating, filtering, and representing information on various devices and displays (e.g., handheld devices, laptop computers); locational dependencies (i.e., allowing different things to happen depending on the physical location of a student); interactive learning objects; and an intelligent agent framework. The S3 environment is highly customizable and supports the coordination of people, activities, and materials with real-time sensitivity to inputs from students.

2.9.4 The Need for a Career Framework

Joyce Malyn-Smith contended that for computational thinking to get traction in the K-12 education community, it needs to be connected to frameworks and standards that are already implemented nationwide. An analysis of the Information Technology Career Cluster Initative's model, for example, provides a way to organize a hierarchy of skills and knowledge that can be repurposed to support the integration of computational thinking in the K-12 arena. At the most basic level, this information technology skills framework calls for literacy and the ability to use common technology applications. Further up the hierarchy is fluency with information technology, which involves core knowledge and skill sets of technology-enabled workers employed in any industry sector. At the highest level of this model are the skill sets necessary for information technology producer or developer careers—those that involve the design, development, support, and management of hardware, software, multimedia, systems integration, and services. In short, individuals engaged in different activities are likely to need different (though overlapping) sets of technology skills.

3

Committee Member Perspectives

3.1 ALFRED AHO

Committee member Alfred Aho, a professor of computer science at Columbia University, commented on several topics: motivations for computational thinking in education, potential pitfalls in ineffectively teaching computational thinking, the need for investment in infrastructure and tools to facilitate learning of computational thinking, and the role of assessment.

Motivation for Computational Thinking

Echoing the sentiments of Matthew Stone, Aho described three common motivations for explicitly introducing computational thinking into education. First, he argued, computational thinking has an impact on virtually every area of human endeavor, as illustrated by the first workshop report's discussion of computational thinking applications in fields as diverse as law, medicine, archeology, journalism, and biology.

Second, he noted dangers in computational thinking done badly. He recounted a story—"A number of years ago when I was doing some consulting for NASA, I came to Washington and noticed an article in the *Washington Post* that said global warming wasn't as bad as scientists feared because the empirical measure of the rate of rise of Earth's oceans wasn't as bad as the computer models had predicted. It turned out to be a software error. So if we're going into this world of modeling and simulation, I would like to put in a plea for good software engineering practices and making sure not only that the data are correct but also that the programs are correct." To underscore this point, Aho cited an article in *Nature* about bad software in computational science.[1]

Third, Aho suggested that computational thinking plays an important role in developing new and improved ways of creating, understanding, and manipulating representations—representations that can change, sometimes dramatically, the way in which people see problems.

Humanization of Computational Thinking

Aho observed that a number of workshop participants pointed to the humanizing effect of computational thinking. Recalling Idit Caperton's thoughts that using information technology in an appropriate manner "engages people, engages their souls, their passion, and their productivity, and people care," Aho described similar experiences in working with undergraduates. He found that using creative programming projects to hone and develop computational thinking skills motivated students to pursue further education in computer science. Aho described classes in which students work in small teams to create their own innovative programming language and then to build a compiler for it, and he reported that "often the students say the most important things that they learned from this course are not principles of programming languages or compiler design but the interactions that they had with the other students and the fun they had in doing the projects."

Aho also suggested that this kind of response to the use of technology was an effective rebuttal to those who argue that computers and information technology are dehumanizing, as illustrated by Jaron Lanier's arguments in *You Are Not a Gadget*.

Computational Thinking as a Moving Target

Aho acknowledged the community's need for a common definition of computational thinking, development of which is inherently difficult given the rapidly changing world to which computational thinking is often applied. Any static definition of computational thinking would likely be obsolete 10 or 20 years for now, he argued, and thus, "The real challenge for the entire community is to define computational thinking and also to keep it current."

With that thought in mind, Aho stated that he was particularly taken

---

[1] Zeeya Merali, 2010, "Computational Science: . . . Error: . . . Why Scientific Programming Does Not Compute," *Nature* 467(7317):775-777. Available at http://www.nature.com/news/2010/101013/full/467775a.html.

with a point made during Deanna Kuhn's presentation, that computer science and education communities should use computational thinking not just to teach old things but also to teach new things, both new methods and new ideas, to solve new problems, because that's what the people we will be educating are going to be doing in the future.

Need to Apply Learning Science to the Problem of Teaching/Learning Computational Thinking

Echoing Jeannette Wing's original charge for the workshop series, Aho said he also believed that educational theory and developmental psychology would help to inform the teaching of computational thinking regarding what particular content to teach and when to teach it. For example, developmental psychology could help identify the specific concepts of computational thinking that would be most appropriate for young children. More generally, he argued that for computational thinking to be taught effectively, any curriculum for computational thinking should be phased according to a developmental sequence characteristic of the students engaged with that curriculum.

Finally, he also suggested that developmental psychology might have value in contributing to different pedagogical models for learners with different cognitive styles and in shaping the infrastructure and tools needed to teach computational thinking.

Infrastructure for Computational Thinking

Addressing the issue of the infrastructure needed to support a serious educational effort to promote computational thinking broadly, Aho noted that such an infrastructure did not consist only of hardware but also necessarily included continuing funding streams, instruments for gathering data needed to analyze outcomes, and an ongoing data collection effort. He added that the infrastructure would also require ongoing maintenance for and the development of new tools to support computational thinking.

A key element of infrastructure, Aho argued, is the ability to integrate applications. Aho warned that "unless these issues get resolved, we are going to find ourselves in a world of the future which may resemble the software world that we're currently in, which is largely a Tower of Babel, [with] lots of incompatible infrastructures and a lot of expense." This comment prompted Stephen Uzzo to argue that interoperability, access, usability, and portability of data are problems that can be explored through collaboration.

How Do You Know What Students Are Learning?

Aho reflected concerns shared by a number of workshop participants that determining what students are learning in computational thinking activities may be difficult. He noted that assessing how a student has internalized the abstractions of computational thinking may be challenging, and even assessing programming skills can be difficult. For example, he indicated that although program correctness is an essential goal of good programming, a student who writes a correct program (i.e., one that exhibits the appropriate behavior) nevertheless may not have made the conceptual connections that one might expect from someone who has written a correct program. He illustrated this point in commenting on Walter Allan's presentation, in which he observed that "[in thinking about] the kind of thought process that a student is following to get the bunny to eat these carrots, I am not sure what the student is actually learning about some of these much deeper issues that a serious programmer would have to face."

3.2 URI WILENSKY

Committee member Uri Wilensky, a Northwestern University professor and director of the Center for Connected Learning and Computer-Based Modeling, shared his observations on a number of key issues discussed at the workshop, including the motivation and value in teaching computational thinking, the challenges arising from the continuing nonconvergence on one definition of computational thinking, and identification of the best environment and tools for conveying computational thinking to different audiences.

Motivation

Wilensky noted that in recent years, many branches of science and engineering have changed in ways that require researchers to be facile with computational thinking. Disciplines such as biology, physics, mathematics, and so on utilize computational methods to analyze problems and model phenomena.[2] Computational thinking in many ways offers a new way to interact and learn about the world and scientific phenomena. According to Wilensky, in order to effectively engage and contribute to modern science and engineering, future scientists and engineers must be

---

[2] National Research Council, 2010, *Report of a Workshop on the Scope and Nature of Computational Thinking*, Washington, D.C.: The National Academies Press. Available at http://www.nap.edu/catalog.php?record_id=12840. Last accessed February 7, 2011.

able to do computational thinking. Thus early in their academic careers, key computational thinking concepts should be introduced and mastered.

A second reason for encouraging computational thinking is the power that it affords for greater automation of tedious tasks and the ability to manage more complexity for all types of learning and discovery. Mechanical automation allows one to delegate certain tedious tasks and simple problem solving in favor of more complex tasks and problem solving. Indeed, as the problems at hand become more complex from a process and computational perspective, increasingly computational tools and abstractions are needed to analyze and understand them.

A third reason is that computational thinking supports the capacity for complex design and simulation. It enables one to naturally create designs within a specific context that do not require access to different kinds of materials because the materials are represented computationally in the form of data and bits. Wilensky also noted that such simulations could be used to inform public debate and discourse about issues of public policy—simulations could be used as modeling tools to explore alternative scenarios for situations in which the interactions and feedback loops among the relevant elements (e.g., resources) are tightly coupled.

Fourth, computational thinking (and computational tools) can enhance self-expression and collaboration, supporting the use of many different forms of expression and the easy sharing of those expressions. The potential for expression and collaboration can be very motivating to many individuals, especially children. Wilensky suggested that the use of computation in art, music, and other kinds of expressive media is underexplored in much of the available research.

As a fifth reason to motivate computational thinking, Wilensky recalled Caperton's argument that educators do not always have to start with kids but rather can focus on those in positions of leadership. That is, Wilensky paraphrased, "If we are thinking about the citizenship value of computational thinking, then it is shortsighted to not pay attention to the people who are actually empowered to make a difference and to try to change the discourse among that group so that they are computationally literate enough to be able to understand this complex world they are being asked to lead."

Last, Wilensky argued that computational thinking, much like the use of Arabic numerals, democratizes access to knowledge. He noted that the significance of Arabic numerals was not that they were essential to multiplication and division (indeed, there were algorithms for multiplying and dividing Roman numerals), but rather that because they were so much less cumbersome, Arabic numerals enabled many more people to perform multiplications and divisions. Wilensky then said, "The claim I was making is that we can now use computational representation—

which similarly affords greater access to knowledge and new knowledge development."

As an example, Wilensky pointed to work he has done with seventh-grade students to use computational thinking and computer modeling to study segregation patterns in Chicago.

> They started by using some of the NetLogo variation models that were based on the work of the Harvard economist Thomas Schelling, who actually won a Nobel Prize for that . . . last year. Schelling, who was a very learned and skilled economist, took many months to build these segregation models by using lots of checkered boards and moving coins around and flipping them back and forth according to determinant rules. He had the basic thinking that was needed to do those models. What he didn't have was tools that could actually do it quickly enough so that he could consider all kinds of alternative scenarios. Now these seventh graders were doing that and they were asking all kinds of questions that pushed well beyond Schelling, like what would happen if there were some Asians that desired only integrated neighborhoods or what would happen if you had many more sets of groups that had different criteria. All those things could be easily explored within the foundational framework—[but] really [were] pretty much impossible without computational thinking and related tools.

Epistemological Diversity Regarding Computational Thinking

Wilensky said that although he saw a lot of common ground on certain aspects of computational thinking among educators and researchers, there were a number of significant areas where workshop participants saw things differently. (This diversity of perspective was also reflected in the first workshop report.) Specifically, he thought that the different ways of understanding computational thinking discussed in the second workshop fell into several categories: ways of seeing and knowing, ways of doing or capacities, a method of inquiry, and ways of collaborating.

He noted that some panelists talked about computational thinking as *ways of seeing and knowing*. For example, he pointed to Robert Tinker's presentation in which Tinker talked about breaking up the world into different simple processes or pieces as a way of seeing the world not just as objects but rather as various informational pieces that can be attached to objects and processes and manipulated. Wilensky said that in this view, computational thinking as ways of seeing and knowing really represents a different way of understanding the world.

Ways of doing and capacities was another conception of computational thinking present in many of the different presentations. According to Wilensky, this view emphasizes the importance of building, designing, and going through various "constructionist" kinds of activities, as

Seymour Papert would call them. In this sense, *ways of doing* includes issues of modeling and thinking using computation as a way of representing the world and being able to experiment and explore alternative scenarios within the simulated world.

Computation as a method of inquiry was interesting to Wilensky because the computer is protean enough to afford users the ability to explore and manipulate all kinds of processes within a small space. Illustrating this view, Wilensky cited John Jungck's presentation in which Jungck talked about the very small range in the evolutionary possibility scale that is actually represented in real creatures. Through computational thinking researchers can create simulated worlds in which one can explore evolutionary trajectories that never actually came into being, creatures that could have evolved but did not.

Last, Wilensky noted that a number of speakers described computational thinking as though it was a *way of collaborating*. These presenters focused on the ways in which collaboration can be extended as a result of computation and computational tools. New ways to connect and form different groups are no longer necessarily limited by geography. Wilensky held that "instead of a spatial model of collaboration, we have this kind of network model of collaboration where there are many different opportunities for synching up, and that capacity is becoming more and more important in our society, and computation is another way to facilitate that."

A Diversity of Venues for Computational Thinking

Represented at the workshop were a number of different perspectives regarding the most effective environment and tools for teaching computational thinking. Wilensky distilled the points of view as those favoring formal curricular learning versus extracurricular learning and those favoring lab-based learning versus in-the-field learning.

The case for making computational thinking a part of the formal school curriculum was made by several speakers. Wilensky pointed to arguments made by Tinker that the right place for computational thinking is in schools, specifically within the science curriculum, because science already uses computational thinking and computers in major ways. With computational thinking, educators can facilitate all kinds of modeling activities in science that really represent ways of actually doing real-world science as opposed to just sort of learning about science. Wilensky argued that social science research may also be a fertile ground for computational thinking, saying that "social science is another very fertile area to integrate computational thinking because tools now enable us to be able to mine large data sets or to create models that were not possible." The constructs of new representational infrastructure and meta-representational capacities in computational thinking offer possibilities for substantial advances in social science.

Others argued that computational thinking should be its own subject within the formal curriculum. Wilensky pointed to Caperton's presentation, which demonstrated that educators can actually design a curriculum around computational thinking. Tinker and others did note a concern that the fact of already packed school curricula may generate some push back on the idea of computational thinking taught as a separate subject. To this point, Wilensky responded that this may be more a strategic discussion rather than a pedagogical one. Work presented by Paulo Blikstein and others shows that there is room even in current science curricula to introduce computational thinking concepts in a way that they fit, and also mutually support learning of other complex concepts. Others further argue that computational thinking fits best in an extracurricular context. Wilensky argued that each option should be explored.

Lab-based approaches were discussed, as were in-the-field approaches. Wilensky argued that this theme is an important one because it reflects the fact that the public in general and educators in particular "tend to think of computing as these kinds of things that are built into our computers and we tend to do them inside. But there were at least some hints of capacities beyond that." Wilensky pointed to presenters Tinker, Jungck, and Uzzo, whose presentations discussed the use of various probes and sensors in the field to collect data for computational learners to analyze and manipulate. Wilensky stated that these options illustrate that "we are not necessarily limited by this [indoor] model of what computation is. Instead we can think of ubiquitous computation and all the different kinds of ways in which we can do things." Thus in-the-field approaches to computational thinking education must be explored, just as lab-based approaches must be developed.

Speaking for himself, Wilensky argued that computational thinking is important enough that it should not have to be squeezed in on the margins or sneaked in on the side. He acknowledged the pragmatic benefits of such an approach but noted that it is perhaps inconsistent with a serious view of computational thinking as a major new mode of thinking that can be powerful for everybody, not just for an elite few.

Wilensky also believed that it is sometimes a red herring to assert that there is no room in the standard curriculum to accommodate a serious examination of computational thinking. Indeed, he argued, sometimes important ideas in computational thinking can be introduced incrementally along with standard content in a way that makes the standard content easier to learn (and vice versa).

Different Tools for Computational Thinking

Wilensky indicated that another theme emphasized in the discussion was what kinds of tools are being used to enable computational thinking. The workshop revealed a range of approaches in use, many of which were related, although Wilensky noted that the main distinguishing factor in these approaches was the question of whether the tools were developed with a target audience in mind. Thus the question is whether, even with children, educators want to use these specially designed learning tools or whether professional tools should be used. For example, Wilensky proposed that maybe tools such as Scratch are designed more with an audience of children in mind—for a target audience especially tuned to learning and motivation through things such as games. That is, using games can be a major motivational tool for an audience of children learning computational thinking.

Other tools may have been designed specifically to target a professional community. Caperton argued for use of professional-level tools such as Flash, one of the most prevalently used animations programs in the world, in computational thinking activities because this use of authentic tools can be a kind of motivation for students to continue learning. Wilensky pointed to modeling tools such as NetLogo, AgentSheets, and many others that particularly help in science. He reiterated that presenter John Jungck demonstrated the "extent [to which] biology has changed dramatically as a result of computation and all the different kinds of tools that are now in the regular toolbox of biologists that just were not there several decades ago." He went on to say that "these tools have changed what the discipline is and made the science of biology much less a natural descriptive one and much more one that involves modeling and analysis with very large sets of data, for example," suggesting that students interested in pursuing careers in biology may be motivated to learn computational thinking as a result of having access to authentic professional biology research tools.

Create and Modify as Complementary Approaches

Wilensky noted the difference between students writing programs starting from a blank screen versus students modifying existing programs, but argued that both approaches have value in conveying concepts of computational thinking. However, he did caution that the canonical "use-modify-create" sequence is not the only viable approach to teaching the skills of computational thinking. In his words,

> It could be in the very first class that kids might create something, like it might be in a biology class where we might say, "Start with some kind

of creature and give some rule of birthing and dying and see what happens to the system." That's a very small creation, but nonetheless it's a creation, and there will be a diversity of different possible choices that people will come up with, and a comparison of those can lead to lots of insights. So we can think of "creating" in small bites as well, and sometimes creation is a lot easier than modifying as a different kind of entry point, and all of the outcomes are ones that we want.

An Affective Dimension

Wilensky also noted an affective dimension to some of the presentations. Specifically, many of the participants in the activities that were reported in the workshop had done sophisticated programming work in developing genuinely useful applications but nevertheless did not believe that they were, in fact, programming. Wilensky saw this disconnect between their capabilities and their self-reported assessments as a problem worth addressing, and pointed to the importance of boosting the students' confidence that in fact they can master complex topics.

He further drew an analogy to the teaching of reading—"I am struck by how much effort we as a school system put into reading. It is a really difficult process, yet we think it is so valuable that we invest enormous amounts of resources in it in the schools. I want to think of us as being bold enough to try to make the claim that computational thinking and computational literacies are becoming important enough that we ought to be investing major resources into it."

3.3 YASMIN KAFAI

Committee member Yasmin Kafai is a professor at the University of Pennsylvania Graduate School of Education. Her research focuses on the design and study of tools for learning programming.

Her comments at the workshop focused on ways to articulate and teach computational thinking more effectively. They included a discipline-oriented approach for identifying key facets of computational thinking, a developmental progression approach for teaching, a real-world problem-solving approach for identifying concepts and teaching, and a cycle approach (use-modify-create) for teaching and assessing learning.

A discipline-oriented approach, Kafai said, means starting from individual disciplines to identify important and useful aspects of computational thinking. This approach may allow the community to articulate more clearly what computational thinking is and what it is not. Kafai noted, "It is within the disciplines that aspects such as programming, visualization, data management, and manipulation can actually help us illuminate and understand processes."

According to Kafai, the computer science and education communities have not developed what presenter Jill Denner termed a "developmentally appropriate definition of computational thinking." Kafai acknowledged that

> [w]e all have examples of kids of many ages and adults who are being very courageous and interested in doing computational thinking, but we also know from prior experience in mathematics and science education that we really do need a more profound understanding of what kids' engagement with computational thinking at different ages is, and then how we can kind of build pedagogies, examples, on it. I think we are far away from that point. These presentations here gave us some ideas about where to start looking and where the examples are.

Kafai found the approach presented by Danny Edelson and Robert Panoff helpful. They focused on how computational thinking can help ask interesting questions and solve real-world problems, rather than simply develop algorithms. They used computational thinking to help students answer questions such as, What's real? Where are the issues? Where are the anomalies and what do they mean? Kafai argued that these questions point to a "kind of social aspect of computational thinking which we don't talk enough about but which would be really important in the social relevance of bringing computational thinking into the disciplines and judging what the value is."

Kafai is a fan of the cycle approach to learning and teaching computational thinking; she argued that the workshop's presentations seemed to come together in favor of this approach as well. "I think we have some convergence here on a kind of cycle approach," Kafai said, "and I know other presenters before us also alluded to this kind of use-modify-create as a kind of pedagogy to introduce students into approaches to computational thinking."

Kafai added, "I don't think it's so bad that the kids get some pieces of code to start with, rather than . . . a blank screen and . . . [the expectation that they] develop all the programming on their own, especially if they don't have any prior competencies in it." She argued that learning to use the code and manipulate it is a good way to try out strategies before designing one's own programs. In addition, the cycle approach works across the disciplines and can be used to facilitate computational learning based in data analysis, visualization, and game design approaches to teaching computational thinking. Kafai felt that the next step was to articulate some extensions and caveats to the cycle approach in order to build better assessment tools.

3.4 MARCIA LINN

Marcia Linn's comments focused on the value and trajectory of computational thinking and on the challenges associated with incorporating computational thinking into the curriculum. Linn, of the University of California, Berkeley, believes that computational thinking is important for everyone for many reasons, including:

- Making personally relevant decisions as a citizen in a scientific and technologically advancing society,
- Succeeding in a growing number of disciplines and jobs,
- Increasing interest in the information technology professions, and
- Enhancing U.S. economic competitiveness in the international sphere.

Linn echoed a point made by presenter Taylor Martin—that computational thinking empowers learners. When learners successfully combine disciplinary knowledge and computational methods they develop their identity as STEM learners. These opportunities for empowerment and expression can affect the way students think about themselves and their potential for continuing in STEM fields.

Linn argued that computational thinking is a powerful concept that by its very nature involves multiple disciplines. She recommended characterizing the trajectory of computational thinking from elementary to college courses. In her view, computational thinking has a role in nearly every discipline and at every level of learning.

Linn acknowledged that the community has not settled on one consensus concept of computational thinking. She felt that recent work—including the two National Research Council workshops on the topic of computational thinking—has resulted in a growing set of compelling examples and some emerging criteria.

The examples characterize the types of reasoning and disciplinary problems that could illustrate computational thinking at every grade level and for a wide range of courses. Several such examples follow:

- *Human genome sequencing.* To understand human genome sequencing, learners need to combine computational ideas with disciplinary knowledge about genetics. The computational ideas that students need to integrate include those of repeated applied algorithms; precisely formulated, unambiguous procedures; search, pattern matching, and iterative refinement; and randomization as an asset in repeated fragmentation. The disciplinary knowledge includes the notion of DNA as a long string of base pairs.

- *Modeling of economic or sociological systems.* To understand modeling of economic or sociological systems, students need to combine computational ideas with disciplinary knowledge of economics and sociology. As an example in economics, consider the idea of aggregating multiple independently specified rule-based agents and sensitivity to initial conditions. As an example in sociology, consider knowledge of community as a collection of independent decision makers.

Linn argued that the criteria for identifying computational thinking are emerging. They include combining computational ideas with disciplinary knowledge. Successful applications of computational thinking involve a process of design. Learners who do computational thinking engage in a sustained process of investigation that results in a novel solution to a problem.

Linn thinks that the value of computational thinking lies in its ubiquity, contemporary role in scientific research, and potential to motivate learners. As the character and criteria for computational thinking are refined, it will grow in importance in the curriculum.

Linn argued that incorporating computational thinking into the curriculum, especially for precollege learners, faces many challenges. She pointed out that it is still not clear whether computational thinking should ultimately be incorporated into education as a general subject, a discipline-specific topic, or a multidisciplinary topic. She noted the conundrum that the goal of becoming literate in computational thinking may not be achieved by taking a course on computational thinking but rather by studying topics in various disciplines that require computational thinking. Indeed, it may be necessary to study computational thinking in several disciplines to fully understand its scope and nature. Only by exploring computational thinking in multiple disciplines can learners appreciate its common features and the challenges of using computational thinking in a new discipline.

Computational thinking is emerging in new specialties that integrate disciplinary knowledge and computational algorithms. Linn argued that computational thinking would be most effective when integrated into specific disciplines rather than as a stand-alone course. Linn remarked, "It seems more efficient to take a disciplinary course and create activities that use computational ideas to advance understanding, but the case could be made for other solutions."

She noted the challenges associated with incorporating computational thinking into the already-packed school curriculum. Computational thinking activities require access to technology, development of new curriculum materials that align with standards, teacher professional development, and building of a community of users who can try out and refine the activities.

Understanding where and how computational thinking fits into current courses will require a concerted effort. Linn remarked, "We could call for emphasizing computational thinking everywhere and end up finding that it is nowhere because no one felt responsible for it. In addition, even if we did incorporate computational thinking into every course we might fail to build competence because the experiences were not cumulative. We need to think about ways to build coherent understanding of computational thinking as students encounter it across disciplines." Linn saw the overarching goal of the workshop as being to catalyze thought about the steps needed to make computational thinking central to all of education.

Linn also commented on the nature of the curricular materials available for teaching computational thinking. Although materials exist, they are not widely available to educators and may be optimized for home use. The available materials generally result from small-scale grassroots efforts or are centered on technology environments. For example, many students are using computational learning tools outside the classroom (such as Scratch or Alice) but do not see any connection between these tools and what they may be doing in school. Teachers could use these tools to enable students to combine disciplinary knowledge and computational thinking.

Many of the available materials are designed to encourage students to explore complex problems fairly autonomously. They need trial and refinement to meet the needs of a broader audience. It is not clear how to make these materials useful and available throughout the educational system.

To make computational thinking a central part of the curriculum in all the relevant courses, Linn argued, requires making the case that it is essential to each discipline. To convince the K-12 community that computational thinking is central requires proactive work with teachers, school administrators, and policy makers. Linn recalled Christine Cunningham's comments about the importance of beginning with teachers and administrators to persuade them of the value of incorporating computational thinking into the curriculum. She also related her own experience that getting school administrators on board "has made an enormous difference in creating a willingness to sustain the use of technology-enhanced materials and even to obtain resources dedicated to using those materials consistently."

Linn drew an analogy to the challenges that exist with respect to adding projects to the STEM subjects. In her view, K-12 students should do at least one 2-week project every year. Such a project would be a natural venue for using computational thinking. She argued that advocating for allocating time in the curriculum to projects—time that would be used to support the teaching of computational thinking and other STEM subjects—would be effort well spent. Teachers could select projects and appropriate technologies. They could use any technology including paper and pencil but would have an opportunity to use powerful computational tools.

Linn noted that the project-based format is particularly well suited for computational thinking because it allows for the kind of sustained reasoning and iterative refinement that occurs when a student is doing a complex task. By contrast, most K-12 curricula do not require students in STEM courses to engage in sustained activities.

She also argued that this type of effort will require the formation of a community of teachers who support each other and mentor newcomers. A one-time summer workshop will not be sufficient. Computational thinking education cannot succeed in the long term without several teachers at every school doing the same thing, "because if you don't have a community, you don't have anything that can sustain this kind of exciting, innovative work."

As far as options to start this integration, there was quite a bit of discussion at the workshop as to whether the best initial approach would be to start with the informal extracurricular activities or the typical school curricula to incorporate computational thinking. Linn suggested targeting both approaches since each will reach a different learning audience. "We are always going to be reaching kind of a different population starting in after-school and summer programs than if we start in school. . . . We know that these after-school and summer programs reach a wide range and often a very deserving group of students, but there are many students that just never have that opportunity." Instead, she argued that the goal should be to target a large audience to maximize the positive empowerment factors of computational thinking.

In summary, Linn saw great potential for computational thinking as a new focus for the curriculum. She was excited about the synergy between course projects and computational thinking. She sees computational thinking as adding motivation for course projects while enticing students into STEM disciplines and preparing them for contemporary careers.

3.5 LARRY SNYDER

Committee member Larry Snyder is a professor of computer science and engineering at the University of Washington. Snyder has researched the topic of fluency with information technology for the past decade.

Throughout the workshop discussion, Snyder expressed particular interest in several key discussion topics: comparisons of programming pedagogies involving programming modification versus novel program creation, opportunities for teacher education and development, and tools and options for integrating computational thinking into the school curriculum. While Snyder felt that assessment of learning is an important topic as well, he argued that in the absence of having a firm definition of computational thinking, "assessing how well we're doing at it is probably a little bit premature." He felt that it is clear that there is a lot to do in this area and continued research is important.

Snyder was interested in the rate of transfer of concepts in computational thinking among students who participate in projects that encourage them to read and modify an existing program versus projects in which students create programs from scratch. To illustrate this issue, Snyder cited points made by Jill Denner and Paulo Blikstein. A result of projects in which a student modifies programs prior to creating them can sometimes be that the student "stalls" at the "modify" stage and never advances to the "create" stage.

Snyder agreed that the use-modify-create model is an excellent way to formulate the computational learning challenge, but he also felt that it is important to understand and assess a student's overall progress rather than focus on at which stage the student gets stuck. Denner agreed, replying that much more research is needed before she can be sure, but she believes that determining a student's overall progress may be attributed to an aspect of "intrepid exploration, a willingness and a confidence to confront the complexity and not back away from it when they're confronted with something that's difficult." Denner added that nearly every student has some ideas that could be executed using computational thinking, "but students come in with different levels of comfort with engaging and with going through the process of trying something, failing, trying something, and failing."

Snyder recalled that Blikstein had a different perspective. Blikstein argued that it may be dangerous to assume that models that seem to go from simple to complex—such as modifying a program first, and then creating a new one from scratch—offer pedagogical benefits. Snyder acknowledged this point, noting that young learners are capable of creating programs, even before they can read programs from other people; in fact, students likely prefer to create their own original programs, which in turn may motivate them to learn more computational thinking skills and concepts.

Snyder made several points highlighting some of the different approaches available for teacher education and development related to computational thinking. Presenter Michelle Williams mentioned several options she and her colleagues have explored, including summer and multiweek programs, in-service training, and so on. Snyder agreed with Williams that teacher development is critical in getting teachers up to speed and prepared to teach computational thinking. Snyder was particularly struck by the concept of teachers learning through working directly on computational thinking-related projects and activities, much like their prospective students would, rather than through rote lecture.

As far as facilitating integration of computational thinking into the school curriculum, Snyder reiterated some of the practices put forth in Jeri Erickson and Walter Allan's presentation. Practices such as teaching teams in which several instructors in related subjects collaborate to instruct a group of students, introducing technology for the project a year in advance, and making the software available on every laptop throughout the school system are a few examples Snyder believes could really increase dissemination of computational thinking. Snyder also appreciated the use of both the technology-based computer game and the physical grid-mapped tarp to make the computational programming concepts as well as the ecology concepts behind the bunny foraging project more concrete.

3.6 JANET KOLODNER

Janet Kolodner is a professor of computing at Georgia Institute of Technology. Her research focuses on the cognitive sciences and learning sciences, and the roles of computing technologies in promoting and mediating learning.

Overall, Kolodner found this second workshop to be much more grounded than the first in learning research and in-the-classroom feedback. Rather than talking in the abstract about what computational thinking might be, discussion focused on real examples of the use and promotion of computational thinking, as well as cases where computational thinking may not have been furthered. Kolodner found that some of the discussions delved very deeply into many practical issues associated with developing computational thinking curricula. In particular, she found presentations by Robert Tinker, Mitch Resnick, and Robert Panoff particularly impressive in this respect.

She pointed out that the first two panels had helped add maturity and depth to her understanding of what computational thinking is, and that some of the later discussions had helped her further refine and develop aspects of this conception of computational thinking. This was particularly impressive given that Kolodner, through her work with the committee and similar computational thinking activities, already had a sophisticated understanding of computational thinking and its attendant issues. Particularly valuable to her were contributions regarding what educators have done to help kids learn to become computational thinkers, and the ways educators might integrate those things into the curriculum.

Kolodner's comments focused on several themes:

- The need for a formal definition of computational thinking,
- Two dueling definitions of computational thinking and their relationship to each other,
  - Pedagogy and learning progressions explored in the workshop,
  - Pedagogy and its role in assessment,
  - Targeting specific goals for assessment,
  - Distinguishing learning assessment from project evaluation, and
  - Setting standards and baselines for assessment.

Definitions of Computational Thinking

Kolodner argued that the computational thinking community needs to be able to identify exactly what is meant by computational thinking to decide what learners should learn and to assess and evaluate what learners know, what they can do, and their attitudes and capabilities with respect to computational thinking. The community must be specific about the definition of computational thinking. The multiple definitions collected at the first workshop are a good start to the discussion but not enough on which to base assessment tools. She noted, "Interestingly, I am left with two not-quite-consistent views of what computational thinking is and what everyone should be capable of. Furthermore, I think this tension is something that warrants further discussion."

Kolodner noted that the first view of computational thinking was described in the presentation by Tinker and later elaborated on by Danny Edelson. Tinker defined computational thinking as fundamentally about breaking problems into smaller and smaller problems that are solvable by rather simplistic computational devices. Edelson, in his discussion, talked about the fact that those who are AI (artificial intelligence) experts and are knowledgeable in computational modeling of cognition have learned to describe mental processes well enough so that they could be run on a computer. Further, this community of experts has a disposition toward describing various processes in that amount of detail in anticipation of using computers to assist in determining solutions. Edelson claimed that computational thinkers aim toward solutions that are constrained by the machine and aim toward breaking problems into parts or chunks that make sense computationally. Making sense computationally means that one can specify the sequencing or control within each chunk, and the chunks (or many of them) each have a particular function. Kolodner stated, "Some of us who come from computer science, and especially cognitive-type AI, are really good at breaking down problems to a set of functional components—the pieces each play useful roles as part of a process, and they can be fit together in a variety of ways to create other processes that perform bigger functions. It's where we feel comfortable, and we can never understand why someone else would break a problem down any other way (but people do—how odd)." Kolodner found this characterization effective and useful in describing what computational thinkers do.

Kolodner pointed out that this definition of computational thinking is far more constrained than simply thinking about computational thinking as problem solving. Rather, this definition regards computational thinking as "a certain kind of problem solving that computer scientists are pretty good at," in particular, thinking in terms of processes to be carried out, imagining the functional pieces of those processes, and identifying which of those pieces have been used in solving previous problems and which might be used in solving later ones.

Notice that this approach is not synonymous with programming. In fact, Kolodner pointed to the work of Richard Lipton of the Georgia Institute of Technology, in which he and several colleagues figured out a treatment for the AIDS virus in patients by mapping out the biological processes within a person's body, the substances those processes use and create, the conditions under which they work that way, and how the processes are sequenced, and then identifying ways in which the sequence of processes might be changed or disrupted. In this way, he used a computational approach to address the problem, but without programming.

This view of computational thinking is consistent with systems thinking and with model-based reasoning, both of which play a huge role both in scientific reasoning and in engaging in computational sciences. Indeed, both Tinker and Panoff proposed integrating model building, simulation, and model-based reasoning into math and science classes as a way to engage kids in computational thinking as they are getting to greater understanding and raising and solving problems in mathematical and scientific domains.

Kolodner added that she believes that computational thinking is a set of skills that transfers across disciplinary domains. She compared computational thinking to the processes involved in inquiry, noting that just as inquiry is not one specific skill but rather a collection of relevant skills specialized for different disciplines, so too is computational thinking a collection of skills that may be applied differently to different disciplines. As an example, Kolodner stated, "If you are a chemist, you are paying attention to different things than if you are a physicist or a biologist, and you answer questions by different means. You might use experimental methods or modeling methods or simulation methods or data-mining methods as you investigate. But in all sciences, you are, in general, attempting to explain phenomena and collecting evidence to help you answer questions about those phenomena and develop well-formed explanations." She believes computational thinking may or may not include quantitative elements, but it always includes, in some way, manipulation of variables, decisions about selecting "the right" representations, and decomposition of complex tasks into manageable subtasks, to name a few.

Although Kolodner is partial to the problem-solving view of computational thinking just described, she was also drawn to a second view of computational thinking put forth by Mitch Resnick. In his view, computational thinking is not simply for problem solving. Rather, he believes that for most people, computational thinking means expressing oneself by utilizing computation fluently. For Resnick, computation's power is in allowing people (everybody, not just those who are good problem solvers) to express themselves through a variety of media. In this view, computational thinking means being able to create, build, and invent presentations and representations *using computation*. This requires fluency with computational media.

Relationship Between Two Views of Computational Thinking

Kolodner argued that a deep understanding of computational thinking may encompass a synthesis of these two views. She synthesized the Tinker/Edelson view and the Resnick view as follows:

> Computational thinking is a kind of reasoning in which one breaks problems/goals/challenges into smaller pieces that are doable by a stupid computational device. This, in general, means thinking in terms of functions that need to be carried out to achieve a goal or solve a problem (not functions in the mathematical sense, but rather in terms of how things work) and pulling apart those problems/goals/challenges into smaller pieces that are functionally separate from each other and where the functions that are pulled out tend to repeat over many different situations. Computational thinkers tend to break problems into functional pieces that have meaning beyond the particular situation in which they are being used. These functional pieces can then be called on repeatedly in solving the problem or combined in new ways to solve new problems and achieve new goals and challenges.

Resnick's view of computational thinking comes into play when one thinks about the role the computer might play in helping to break problems into pieces and compose the pieces in new ways. To the extent that the computer can help with this kind of thinking, we become capable of achieving bigger goals or solving more complex problems. But this requires two things: (1) that we develop tools to help people think computationally (e.g., one could think about Scratch in this way) and (2) that we be able to use those tools fluently. A computational thinker is fluent in this kind of thinking and in using some set of tools that help with this kind of thinking.

With respect to computational thinking for everyone, the implication is that all individuals should get as far as being able to use these types of tools well to help them solve problems, meet challenges, or express themselves. Some will become more proficient, being able to manipulate these tools and solutions to create, build, or invent better solutions or creations. At the highest level are those who will be able to use computational media and thinking in the most sophisticated ways—as scientists, computationalists, and even artists.

Yet, the relationship between these two views of computational thinking is not entirely clear, and there may be a certain tension between the two. Certainly, Kolodner argued, there is overlap, for example, for those whose expression is of sophisticated complex systems. Those learning to be computational biologists and computational physicists and so on might need to have capabilities in both domains of computational thinking: problem solving/modeling and expression. But beyond this point, the relationship between the two characterizations of computational thinking is not clear. It is not clear that beginning with developing capabilities within the realm of View 2 (expression) is necessarily the way to get students to develop capabilities within the realm of View 1 (problem solving/modeling). Similarly, it is also not clear whether those who are facile at the skills and practices of View 1 will automatically be facile at the skills and practices of View 2. Kolodner believes this blurred relationship is "a really interesting conundrum that needs more attention from the research community."

Helping People Learn to Be Computational Thinkers

Presenter Derek Briggs of the University of Colorado, Boulder, put forth a question during one of the panel discussions that Kolodner found helpful in articulating how to promote computational thinking. Briggs questioned the goals sought with respect to learning computational thinking. He wondered whether we want to be able solely to build tools that will help people reason better computationally, or rather whether we believe that computational thinking is something we want everybody to learn. He pointed out that if the latter is the case, then we seem to be going against the grain, because we know from the learning sciences and from education best practice that it is hard to learn skills disembodied from the contexts in which they are used.

Kolodner argued that the community has both goals—tool building for better computational thinking and computational thinking as a core skill for everyone—and that Briggs's warning about teaching computational thinking in context is a key reminder of best practice. She went on to say that the education community should most definitely be aiming toward helping everybody learn computational thinking and that, yes, the community does seek to promote computational thinking as a set of necessary general-purpose skills. Kolodner believes it is important not to fall prey to the mistaken notion that if one learns computational thinking skills in one context, one will automatically be able to use them in another context. Rather, it will be important to remember that one can learn to use computational thinking skills across contexts only if (1) the skills are practiced across contexts, (2) their use is identified and articulated in each context, (3) their use is compared and contrasted across situations, and (4) learners are pushed to anticipate other situations in which they might use the same skills (and how they would).

These four guidelines come from the transfer literature—the chapter on transfer in *How People Learn*[3] makes them clear. Kolodner pointed out that following these guidelines is absolutely necessary in designing instruction—otherwise, we are only helping kids learn to program or learn to use some set of skills in some particular contexts. This is analogous, she added, to what we now understand about learning to be a scientific reasoner. Scientific reasoning, or inquiry, is not a simple skill that one learns in one domain and applies in a bunch of others. Rather, scientific reasoning is a set of complex cognitive skills that one must learn to carry out flexibly over a variety of domains, and the way to help kids learn that is to help them carry out scientific reasoning over a variety of situations, help them recognize what they are doing, and help them recognize how their reasoning is similar and different over a variety of situations. The workshop touched on these issues in the discussion, but the four guidelines were not entirely articulated.

This set of guidelines is really important for educators to remember with respect to computational thinking; if kids are introduced to computational thinking only in the context of programming and never think about how to use computational thinking, or never have opportunities to use computational thinking in other situations, then they may not develop computational thinking. Mike Clancy's cases are interesting with respect to this—they make the computational thinking of experts visible as a way to illustrate computational thinking applied to a domain. Kolodner wondered to what extent students who use those cases take their compu-

---

[3] NRC, 2000, "Learning and Transfer," in *How People Learn: Brain, Mind, Experience, and School: Expanded Edition*, Washington, D.C.: The National Academies Press. Available at http://www.nap.edu/catalog.php?record_id=9853. Last accessed May 20, 2011.

tational thinking outside the computer science class, and what it would take to promote that type of cross-domain application.

Several people, across both views of what computational thinking is, talked about teaching computational thinking concepts and skills through a learning progression paradigm of use, modify, and create. Kolodner thought that many of the examples of computational thinking learning discussed in the workshop reflected adoption of this approach to teaching computational thinking, with varying levels of success.

One example was Tinker's learning progression for learning computational thinking in a science class, learning that involved the following:

- Numbers are associated with things and their interactions (e.g., temperature),
- Values change over time,
- Changes can be modeled,
- Models involve lots of little steps defined by simple rules (e.g., molecular dance),
- Models can be tested to find a range of applicability,
- You can make models, and
- Many applications of computers share these features.

If using models is done repeatedly in science classes, and if kids gradually move from using to modifying to creating their own models, and if they discuss the features behind the models—why they are the way they are, why and how one might want to change them, and how they went about making changes and creating new models—then there is a good chance that kids will learn to think fluently about running, trusting the results of, revising, and maybe, designing computational models. If, in addition, they discuss how what they are doing is similar to what computer programmers do and/or how it is similar to other problem solving and design, they will broaden their understanding and capabilities with respect to computational thinking. Kolodner added, "The deal is that one develops the ability to broadly use cognitive skills to the extent that one has experiences using them in a variety of situations, considers how one was using them, and anticipates their use in other situations." So, for example, one could start from science class and broaden out from there. Edelson's analogy between computational thinking and geographic computational reasoning illustrated this point. If one helps kids reason geographically, helps them see that process as computational reasoning, and helps them anticipate other ways that reasoning might be useful, one can use that as a base and broaden knowledge and use of computational thinking from there.

Kolodner was very interested in perspectives on learning progressions associated with older children. Specifically she wanted to understand at what point students were capable of creating their own computational models using computer programs rather than just using existing models and manipulating them. She noted that around middle school age, students seem able to grasp increasingly sophisticated computational and programming concepts. This observation seems consistent with a point presenter Tinker made that at around fourth grade seems to be when a number of factors such as student development, teaching resources, and opportunity converge and make computational modeling more likely. Tinker also added that creating a computational model from scratch on a computer can require a great deal of time learning programming to realize that model. On the other hand, systems like NetLogo and AgentSheets allow students to manipulate models someone else built without necessarily having to master a whole lot of detail themselves, and then allow looking inside those models and changing them before beginning from scratch to build one's own models.

Presenter Christina Schwarz added some warnings to this discussion, pointing out again that one cannot just assume that students will learn computational thinking through model building. She pointed out that instructors have to be realistic about students and their motivations to build models. When projects have them focus on concepts that they already understand based on outside or prior knowledge, students may be more likely to explore and try more complex models. If concepts are brand new, however, students need to explore before they can do complex model building. And they certainly won't be able to learn new computational thinking skills or concepts while they are struggling with some new science concept.

Kolodner agreed and emphasized that those creating curricula should be sure to think longitudinally—the focus should be on creating more opportunities to model year after year, helping learners to gradually build up their ability to model and their computational thinking capabilities. Their progress on both should be tracked over time. She also highlighted one more important caveat about the use and promotion of computational thinking in the classroom: simply programming, or even simply teaching students to program, is not necessarily promoting computational thinking.

Kolodner expressed concern over a thread of discussion running through some of the presentations that seemed to presume that as a part of the process of learning to program, students would learn computational thinking. For Kolodner, a big question is how an instructor can be sure that students engaging in programming activities are actually learning computational thinking. Similarly, do students themselves realize they are learning thinking skills that can be applied outside the constraints of the particular activity they are engaging in? Or are the students just becoming better programmers or model builders or game players?

To get a clear picture of what is happening in a computational activity in terms of assessment and evaluation, one has to apply an entire toolbox of assessment and evaluation tools, according to Kolodner. One tool or method is not enough. Kolodner believes that a student's reflecting on a computational activity, being able to teach or help someone else learn the concepts, or being able to effectively articulate the relevant computational process at issue can be seen as likely indications that the student is learning computational thinking. As students are able to use increasingly elegant, efficient, and sophisticated approaches to tackle computational thinking tasks, this ability can also demonstrate learning and improvement in computational thinking, Kolodner believes.

Another important point is that one cannot presume that just because one is programming, one is learning to be a computational thinker. Kolodner pointed out the importance of remembering that separating out the abstract processes from the specifics of what one is doing does not come easily to everyone. Referring back to points from *How People Learn*,[4] she stated that to learn computational thinking from programming experiences, learners need to engage in thinking about what they are doing and under what other circumstances they might use the same type of thinking. Also, she was concerned that perhaps this assumption (that learning to be a computational thinker would arise simply from learning to program) reflected confusion over what computational thinking is. Although programming may be one tool that is used to teach or highlight computational concepts, it is not synonymous with computational thinking, and Kolodner again warned that a good definition of computational thinking is needed—both so that curricula will be designed to promote computational thinking and so that achieving capability in computational thinking can be measured well.

Pedagogy as a Criterion for Assessment: An Elegant Relationship

Kolodner believes that assessment and pedagogy can be rather elegantly related to each other. She pointed to arguments from Clancy and Blikstein, who both talked about pedagogy as a lead-in to assessment. Clancy talked about how the case studies in his lab-centric approach, as well as the derivative pedagogy, provided lots of criteria for assessing how well learners are actually doing computational thinking. In Clancy's

---

[4] NRC, 2000, *How People Learn: Brain, Mind, Experience, and School: Expanded Edition*, Washington, D.C.: The National Academies Press. Available at http://www.nap.edu/catalog.php?record_id=9853. Last accessed May 20, 2011.

approach, learners are learning to program (and could be learning computational thinking) through the use of case studies that show how others have solved similar programming problems. He pointed out that the decisions about what content to put into cases, and then how to evaluate and assess learners' computational thinking, go hand in hand with each other. Blikstein talked about animated representations students develop in his activity and how when combined with the underlying pedagogy of the activity, analysis of the drawings allows certain kinds of assessments and ways of interpreting what the kids are saying and doing.

Goals of Assessment

In addition to knowing what one wants to assess, one must consider the purpose of the assessment, because the reason for any assessment plays a critical role in determining the data and process necessary to perform it. Kolodner identified three reasons for assessing computational thinking: (1) to judge the curriculum and related materials and pedagogy, (2) to judge the progress of individuals, e.g., for giving grades, and (3) to manage instructor training and support. Kolodner noted that the kinds of data relevant to each reason would not necessarily be identical.

Assessment versus Evaluation

Kolodner explained that assessment is not the same as evaluation, although the terms are often used interchangeably. According to her, assessment is about measuring what people have learned, how they feel about something, or their capabilities. Formative assessments deal with discovering what has been learned along the way to inform what comes next. Presenters Jim Slotta and Mike Clancy both noted the importance of capturing some of the reasoning learners are doing that otherwise would be invisible in a formative assessment in order to explore when and how one might change instruction along the way to improve learning. Summative assessment occurs at the end of a module or semester or project to determine how much knowledge was gained overall. Evaluation, on the other hand, speaks more to how well a curriculum or a software tool is working—its efficacy, its costs, its usability, and so on. Kolodner agreed with presenter Cathy Lachapelle of the Museum of Science, Boston, who also discussed evaluation, specifically with respect to the need for usability in a computational thinking project in order to incorporate computational thinking effectively into a curriculum and make it widely available.

In response to discussion from Lachapelle, Kolodner said that the computational thinking community should consider at some point creating its own assessment framework. The National Assessment of Educational Progress currently looks at subjects like science, math, technology education, pre-engineering, and so on, but does not assess computational thinking.

Standards and Tactics for Assessment and Evaluation

Kolodner echoed the sentiments of several presenters (Briggs, Clancy, and Schwarz) that assessment and evaluation are more than just collecting data points. They are about doing comparisons and analyzing outcomes. Sometimes those comparisons are as simple as what a researcher hypothesized versus what actually resulted. Presenter Derek Briggs argued that there must be some standard or baseline to which researchers must compare results. Briggs focused on learning progressions and constructs as one example of a standard or baseline for comparison. Kolodner called the process by which a researcher considers what standards and baselines to use and embeds those standards in the computational thinking project, the "tactics" of assessment used. In some cases, the researcher does not select his or her own baselines or learning progression but instead adopts them from an external source. Kolodner pointed to presenter Christina Schwarz's experience dealing with her local school district's biology learning progression guidelines for middle school students as an example of an external baseline.

Repetition and Reflection as an Assessment Tactic

One tactic Kolodner endorsed was repetition across disciplines combined with reflection. She argued that scientific reasoning and computational thinking should be done in a number of different subjects and repeated over and over in order to help learners understand both the similarities and the differences in the ways in which scientific reasoning and computational thinking are done as well as develop general skills in computational thinking. To cross disciplines effectively, Kolodner argued, there should be some sort of reflection on what it is that has been done as well as some anticipation of other circumstances in which skills and lessons learned would be useful.

Kolodner also felt that reflection on pedagogical content knowledge with respect to computational thinking is important for instructors of computational thinking. In response to Michelle Williams's presentation, Kolodner asked for more information about how the reflection questions were developed that were posed to teachers after they had completed a teacher development computational thinking learning project. In essence, if the purpose of having the teachers complete the same project that their students would do later was to provide scaffolding in a systematic way, Kolodner wanted to understand the underlying system better.

Embedded Assessments and Tracking/Logging Data

Embedded assessments, especially those that capture online the thinking of learners, allow assessment of student understanding that a researcher may not have access to otherwise. Kolodner noted that Briggs talked about collecting performance data and Slotta mentioned the value of real-time reflections on threads of collaborative discussions among the students. They argued, and Kolodner agreed, that these embedded real-time assessments allow "getting in there and really dealing with the issues that the learners are having at the moment that they are having them. Maybe at the moment they are having them, maybe later, but the talking uncovers things that you might not see otherwise."

Kolodner believes that tracking of activities seems particularly important to analyzing computational thinking. Whether Blikstein's log files, or Schwarz's interviewing to help track thinking, or Clancy's noting details of collaborative discussions, such tracking enables particularly important and informative project assessment and evaluation.

Kolodner finds that it is hard to tell who to go to concerning community building in the education community and the various disciplines. She stated that "people seek environments that align to their ways of thinking and working. We all do it, and this self-sorting process tends to create silos." Kolodner argued that such silos will not help computational thinking have a wide impact.

3.7 BRIAN BLAKE

Brian Blake is a professor of computer science at the University of Notre Dame and is associate dean for engineering. His research areas include software engineering and, more recently, methods to make advanced computer science techniques digestible for those who are not in the same specialty. The latter effort is intended to attract underrepresented minorities into computing.

In his comments to the workshop, Blake expressed the evolution of his thoughts on computational thinking through dialog and interaction with various scholars over the course of the two NRC computational thinking workshops. In the first workshop, he explained, the committee sought to characterize computational thinking by first attempting to look for the existence of computational thinking in other fields, in other ways of thinking. From there the committee could then classify and describe it as computational thinking in a way that would enable researchers and educators to re-embed it into training students or retooling teachers or professionals.

Blake went on to explain that his experience over the past year, based on the first workshop and his own personal observations of his son's learning progression from kindergarten to first grade, had caused his thinking to evolve. Now, the notion of developmental milestones is very important to him. He believes that the understanding of computational thinking should be thought of in terms of decomposing computational thinking "elements" into developmental milestones.

Blake noted that during Peter Henderson's presentation on the efforts underway at Shodor, Henderson's example featuring Thomas the Train in solving a routing challenge demonstrates that there seems to be an opportunity to start to understand computational thinking at the lowest levels, and then as we move from K-12 into postsecondary education, we can explore increasing complexity within the milestones.

Blake summarized several main points he had gathered from the second workshop's presentations. There may be an opportunity very early in a child's learning progression to identify significant computational thinking talent. This might be done by looking at specific instances where computational thinking might fold into a learning activity, and then assessing a student's competency with respect to these computational elements. To illustrate, Blake pointed back to Henderson's Thomas the Train example and suggested that a simple activity with embedded computational thinking challenges might be a means of identifying talent. Concerning the idea of training, Blake argued that by taking opportunities to identify and assess computational thinking talent in individual students, and to start to enumerate indicators of such talent, a researcher or an educator might be able to recognize when a student either is demonstrating a significant talent in computational thinking or is at least at the appropriate learning progression level for that age range.

Blake argued that the next step would be to use this process of embedding, assessing, and identifying at the macro level over a longer period of time to identify learning progression baselines. This technique utilizes assessment and evaluation to determine where in learning development a particular baseline is situated.

From the perspective of learning progression at the macro level, the types of concepts to be enumerated so as to identify potentially talented computational thinkers at young ages are not limited solely to concepts related obviously to computer science thinking, math thinking, or even scientific thinking. Instead, these concepts are likely to span all of these types of thinking and analysis. As the emerging computation community moves forward, scholars should perhaps target these sorts of concepts to specify them more clearly and possibly re-embed them for identification of talent and for determination of learning progression.

4

Summaries of Individual Presentations

4.1 COMPUTATIONAL THINKING AND SCIENTIFIC VISUALIZATION

4.1.1 Questions Addressed

- What are the relevant lessons learned and best practices for improving computational thinking in K-12 education?
- What are examples of computational thinking and how, if at all, does computational thinking vary by discipline at the K-12 level?
- What exposures and experiences contribute to developing computational thinking in the disciplines?
- How do computers and programming fit into computational thinking?
- What are plausible paths and activities for teaching the most important computational thinking concepts?

Presenters:
 Robert Tinker, Concord Consortium
 Mitch Resnick, Massachusetts Institute of Technology
 John Jungck, Beloit College, BioQUEST
 Idit Caperton, World Wide Workshop

Committee respondent: *Uri Wilensky*

4.1.2 Robert Tinker, Concord Consortium

The Concord Consortium is a non-profit research and technology development group that focuses on applying technology to improve learning at different grades. Robert Tinker, the founder of the Concord Consortium, argued that computational efforts in K-12 should be integrated around a science focus rather than a focus on either mathematics or engineering.

Elaborating on this argument, he suggested that computational thinking offers an alternative new way of finding out about the world, which is important for citizenship, for future work, and for professionals of all types. Nevertheless, he believes that neither the computer science community nor the education community has yet clearly articulated the essence of computational thinking. As usually presented, computational thinking involves abstractions upon abstractions, which are difficult to make concrete.

At the core of computational thinking, Tinker argued, is the ability to break big problems into smaller problems until one can automate the solutions of those smaller problems for rapid response. (It is for this reason that Tinker believes that engineering is not an appropriate integrating focus for attempts to teach computational thinking—engineering taught at the K-12 level is not particularly amenable to decomposition.) This core, he argued, indicates a possible route for introducing computational thinking into K-12 education.

Tinker's view is that science is the right focus because modern science often uses computational models that are based on scientific principles and whose use depends on visualizations. Understanding these models requires computational thinking—scientific models and visualizations allow students to visualize the computations that are going on in near real time. Tinker noted that students learn better by seeing models and interacting with them, and that by exploring the model in a spirit of inquiry, they learn about the science in the model in much the same way that scientists learn about nature by using the scientific method. He argued that students can learn complicated, deep concepts this way rather than through the more "off-putting" and often confusing approach of formalistic equations.

Tinker proposed an approach across the K-12 curriculum that uses simple models of scientific concepts such as temperature, light, and force to teach computational thinking. A progression of concepts could start in early elementary grades with basic ideas such as "there are numbers associated with things you observe." (See Figure 2.1.) In later grades, students might manipulate and refine models to reflect more sophisticated understanding of the concepts represented in the models. Finally, in high school, a student might be able to select, modify, and apply both hardware and software models as a key part of an extended investigation.

Tinker suggested that the following learning progression could fit into the K-12 curriculum, improve science teaching and learning, and introduce important aspects of computational thinking:

1. There are numeric values associated with every object and their interactions.
2. These values change over time.
3. These changes can be modeled.
4. Models involve lots of simple steps defined by simple rules (e.g., the molecular dance).
5. Models can be tested to find their range of applicability.
6. You can make models.
7. Many other applications of computers share the same features.

When asked whether students perform better when learning through computational modeling and visualization as opposed to a more traditional approach, he replied that such a distinction is not particularly important. Rather than worry whether one method is better than the other, Tinker pointed out that it is a good outcome if a teacher has an additional tool in his or her arsenal to teach a complicated concept.

Tinker noted that because students begin as concrete thinkers, it remains a challenge to identify the age or grade level at which children can handle abstraction. As an example, he said that although he has worked with second graders by hooking up a probe to measure temperature, it is only at fourth grade that students demonstrate reliable results of learning and comprehension with such methods.

According to Tinker, students involved in a very tightly packed K-12 curriculum do not have the time to master programming in order to manipulate models. Rather, he recommends a programming environment such as NetLogo or AgentSheets partially populated with general tools, but still needing interconnection and "tuning," that were designed to focus users on the concepts represented rather than on the details of programming. Another option is to use an existing piece of software in which the student can manipulate important parameters.

4.1.3 Mitch Resnick, Massachusetts Institute of Technology

Mitch Resnick of the MIT Media Lab said that computational thinkers must be able to use computational media to create, build, and invent solutions to problems. He framed this approach in terms of students being able to express themselves and their ideas in computational terms, and emphasized that indeed this should be part of the motivation to learn computational thinking. "When young people learn about language, we don't just teach them linguistics or grammar; we let them express themselves. We want a similar thing with computational thinking."

Moreover, he argued, most people work better on things they care about and that are meaningful to them, and so embedding the study of abstraction in concrete activity helps to make it meaningful and understandable.

Resnick pointed out that for students to express themselves meaningfully with computational media, they need to learn new concepts as well as develop new capacities. He argued that computer science classes often overemphasize computational thinking concepts (such as recursion) at the expense of helping students develop computational thinking capacities for design and social cooperation. Computational concepts include concepts such as conditionals, processes, synchronization, and recursion. Design capacities deal with skills like prototyping, abstracting, modularizing, and debugging. Social-cooperative capacities include sharing, collaborating, remixing, and crowd-sourcing. These social-cooperative capacities are becoming increasingly important as new computing and networking technologies open up new possibilities for widespread cooperation.

Resnick's computational environment of choice for supporting computational expression is Scratch. The MIT Media Lab developed Scratch and a companion online community to help engage people in creative learning experiences and to support the development of computational thinking. Scratch is a graphical programming language, giving the user the ability to build programs by snapping together graphical blocks that control the actions of different dynamic actors on a screen. (Such an approach to program construction enables users to avoid issues of syntax and other details that often distract users from the critical processes of designing, creating, and inventing. Resnick believes this construction process serves an important grounding function for learning abstract computational concepts, making concepts more concrete and understandable.) Scratch also facilitates social cooperation by making it very easy for a user to share his or her design with others for comment and feedback. (See Figure 2.2.)

Resnick provided several examples emphasizing the role of expression through construction and social cooperation from one particular member of the Scratch online community who goes by the username MyRedNeptune. MyRedNeptune was a young student from Moscow and joined the Scratch online community shortly after it went live in 2007. "One of the first projects that she created," Resnick said, "was a type of interactive greeting card for the holidays." Each time a person clicked on one of the reindeer, it would begin playing "We Wish You a Merry Christmas" on a musical instrument in concert with the other animated reindeer. Creating the card required modularization and synchronization, as well as a number of core computational concepts.

Next, MyRedNeptune began offering her consulting services to develop animated characters upon request. Another community member requested that she develop a cheetah for a project. Resnick continued, "She went to the National Geographic website and she found a video of a cheetah. She used that to help guide the graphic of an animation that she developed, and then someone else used her graphic and integrated it into her project."

When yet another community member requested that she show how she developed her animations, she began to develop tutorials in Scratch on how to program animated characters. One of her first tutorials was on how to animate a bird to make its wings flap back and forth. Later she was asked to participate in an international collaboration with five or six other kids in three different countries. Working together, they developed a type of adventure video game, with each child working on different parts of the activity. Resnick noted: "I think you can see from these examples how MyRedNeptune developed as a computational thinker, learning to think creatively, reason systematically, and work collaboratively."

Scratch is used both inside and outside formal school curricula. Initially used in homes, after-school centers, community centers, and museums, it is now moving into the schools and is being used today to teach basic concepts in university-level introductory computer science classes in a number of universities.

Resnick shared that "one thing we've seen is that different kids have different trajectories. Some will spend a lot of time continuing to work on the same types of projects, over and over. You might think that they are stuck, but there's a lot of things happening in their minds, and suddenly they'll start working on new types of projects and ideas."

Resnick and his colleagues are working on many new initiatives to support the development of computational thinking through Scratch, including an online community (called ScratchEd) specifically for teachers who are helping students learn with Scratch.

4.1.4 John Jungck, Beloit College, BioQUEST

John Jungck and his collaborators founded the BioQUEST Curriculum Consortium 24 years ago to bring computation and mathematics into the undergraduate biology curriculum. Jungck noted that although there are many reasons for using computation in biology education, the rationale he presented at the workshop focused on the power of visualization in a biological context. He noted the evolution of paradigms for scientific investigation from empirical (experiment, observation) to theoretical (models, theoretical generalizations) to computational (simulation) to data exploration and e-science (collection of data on a massive scale: exploration facilitated by theory, mathematics, statistics, and computer science). In this context, biology education needs to provide students with ways of understanding biological data—environmental data and genomic data, for example—that is multivariate, multidimensional, and multicausal and that exists at multiple scales in enormous volume (terabytes of data per day).

The philosophy of BioQUEST rests on three pillars:

- Students take an active role in posing problems to examine, much as a scientist has to learn to pose good problems. Good problems must be appealing, have significance, and be feasible to address.
- Students solve problems iteratively. They must learn to appreciate the nature of scientific hypotheses as answers as well as to develop heuristics for achieving closure to scientific problems.
- Students must persuade their peers that a solution is useful and or valid, a process that mirrors the role of publication and extensive peer review in biological research.

The primary challenge for learning in accordance with this philosophy is that in focusing on the student as problem-poser, teachers lose much of the control they traditionally have over the learning process. Students engaging in self-directed collaborative processes may make some teachers uncomfortable. Furthermore, students in this environment may have more technical skill than their teachers, and so peer review from other classmates may be more important than teacher feedback as far as advancing learning.

Jungck briefly described a number of BioQUEST projects. For example, one project sought to develop student facility with the idea of biological modeling with equations. For this purpose, the Biological ESTEEM project (ESTEEM stands for Excel Simulations and Tools for Exploratory Experiential Mathematics) seeks to provide students with a mathematical vocabulary for describing common modeling concepts (e.g., linear, exponential, and logistic growth[1]).

Another BioQUEST project (BEDROCK) focuses on bioinformatics. The BEDROCK project requires students to use a supercomputer tool

---

[1] The Biological ESTEEM Collection, website, BioQUEST Curriculum Consortium, http://bioquest.org/esteem/index.php. Last accessed February 7, 2011.

called Biology Workbench,[2] which allows biologists to search many popular protein and nucleic acid sequence databases. Database searching is integrated with access to a wide variety of analysis and modeling tools. Students can align multiple sequences of a particular gene from different organisms onto one three-dimensional structure and see the evolutionary conservation involved; they can thus relate the comparative biology of sequences to structure, function, and phylogeny.

Yet another project is BIRDD (Beagle Investigations Return with Darwinian Data), whose goal is to provide a variety of resources related to evolutionary research. Labs are rare in courses dealing with evolution, largely because evolutionary phenomena involve temporal and geographic scales that make it difficult for instructors to develop labs comparable to those in biochemistry, physiology, or behavior. BIRDD addresses this problem by providing raw data (e.g., bird songs, sequence data, rainfall, breeding sites, and so on) and pedagogical ideas to help instructors structure appropriate pedagogical experiences for their students. BIRDD helps students generate questions and look at, for instance, whether character displacement happens when the species co-occur or when they inhabit different islands.

To illustrate the special relationship between biology on one hand and mathematics and computation on the other, Jungck noted 10 equations that have driven substantial amounts of biological research and for which numerous educational materials have been developed:[3]

1. Fisher's fundamental theorem of natural selection,
2. Cormack-Hounsfield computer assisted tomography,
3. Genetic mapping (units = morgans; the Haldane function),
4. Fitch-Margoliash little maximum parsimony algorithm (Penny and Hendy—Molecular Phylogenetic Trees—Bioinformatics),
5. Lotka-Volterra interspecific competition logistic equations,
6. Hodgkin-Huxley equations for neural axon membrane potential,
7. Michaelis-Menten equation for enzyme kinetics (Jacob and Monod),
8. Allometry (e.g., MacArthur-Wilson species area law and conservation),
9. Hypothesis testing (e.g., Luria-Delbrück fluctuation test), and
10. Crick-Griffith-Orgel comma-free coding theory.

---

[2] BEDROCK (Bioinformatics Education Dissemination: Reaching Out, Connecting, and Knitting-together), website, BioQUEST Curriculum Consortium, http://bioquest.org/bedrock/about.php. Last accessed February 7, 2011.

[3] John Jungck, 1997, "Ten Equations That Changed Biology: Mathematics in Problem-Solving Biology Curricula," *Bioscience* 23(1):11-36.

He also noted that the typical biology textbook contains only a handful of equations, and even those are linear equations, and expressed his surprise that biological visualizations, important as they are to the way biologists think about the world, are not accompanied by the tools needed to interpret different kinds of multivariate, multidimensional biological data.

Finally, Jungck discussed the Visible Human Explorer (VHE). According to the VHE website,[4] the VHE is an experimental user interface for browsing the National Library of Medicine's (NLM's) Visible Human data set, which is based on two digitized cadavers in the National Institutes of Health Visible Human data set. The interface allows users to browse a miniature Visible Human volume, locate images of interest, and automatically retrieve desired full-resolution images from the NLM archive.

Jungck concluded by noting that computers and computation have transformed biology. He noted a quote from Michael Levitt (a structural biologist at Stanford) that "computers have changed biology forever, even if most biologists don't yet realize it." Educationally, he stressed the work of di Sessa, Parnafes, and others who emphasize the importance of engaging students in constructing, revising, inventing, inspecting, critiquing, and using rich visualizations for promoting conceptual understanding.

4.1.5 Idit Caperton, World Wide Workshop, Globaloria

Idit Caperton described Globaloria as a platform, a transformative social media learning network, with a comprehensive hybrid course (online/in class) for playing and making games. It includes a customizable curriculum, community-developed resources, tools, tutorials, and expert support. Students and educators learn how to create their own web games, produce wikis, publish rich-media blogs, and openly share and exchange ideas, game code, questions, and progress using the latest learning methods and digital communication technology. Globaloria is a project-based learning environment for stimulating computational creativity as well as inventiveness in youth and educators as a necessary skill for the 21st century. Computational projects are built around a range of topics, such as health, climate, alternative energy, civics, mathematics, biology, social studies, and literature.

The World Wide Workshop's innovative R&D and pedagogical approaches to platforms and tools for cultivating computational think-

---

[4] Human Visible Explorer, website, Human Computer Interaction Lab, University of Maryland, http://www.cs.umd.edu/hcil/visible-human/vhe.shtml. Last accessed February 7, 2011.

ing and computational inventiveness have roots in Caperton's MIT and Harvard research, and in educational theories about the value of project-based, multidisciplinary, innovative and creative learning (of any subject) through software design and programming.[5]

Caperton also described Globaloria as a customizable textbook comprising three main units. An introductory or "getting started" unit provides students with the opportunity to establish their own project spaces on the wiki network and to review existing games' operation and their codes. A following unit is "game design," in which students design an original game about a complex topic (in science, math, health, civics) and a social issue that matters to them. Students come up with an idea, assemble teams, do research, build and videotape their paper prototypes, and construct a concept and a demonstration that they present, both physically and online via web conferencing. Using Flash text and drawing and animation techniques, they program an interactive demonstration of their game concepts. A third unit is "game development," in which students develop their game concepts and demonstrations into a complete, interactive game. Each unit contains a structured set of learning topics, as well as projects and assignments structured to help students create critical parts for their own original game.

Globaloria seeks to impart to students six contemporary learning abilities: the ability to imagine, design, prototype, and program an educational game, wiki, or sim; the ability to use project management skills in developing programmable wiki systems in a Web 2.0 environment; the ability to produce animated media, programming, publishing, and distributing interactive purposeful digital media in social networks; the ability to learn in a social constructionist manner and to participate actively in the public exchanges of ideas and artifacts; the ability to undertake information-based learning, search, and exploration as they relate to the abilities above; and the ability to surf websites and use web applications thoughtfully as they relate to the earlier abilities enumerated. Caperton argued that these abilities go beyond the typical media literacy skills, since they emphasize a bundle of complex and sophisticated constructionist digital literacies and involve longer-term engagement (students are required to use Globaloria daily, over two semesters, for a minimum of 100-150 hours[6]).

The Globaloria approach emphasizes constructionist collaboration

---

[5] The canonical examples of such research are Idit Caperton, 1991, *Children Designers*, and Idit Caperton and Seymour Papert, 1991, *Constructionism*, both published by Ablex, Norwood, New Jersey

[6] Caperton recommended repeating the use of Globaloria year after year for greater effects on computational thinking in learners.

within a transparent community. Participants in the community—teachers, students, staff, and game teams—maintain public blogs as design journals, share resources, and publish completed games on the community wiki. They can also submit created games for competitions or for publishing on the school's Globaloria network.

Caperton suggested that it is possible to learn any subject and to master complex topics or social issues by creating functional, representational, educational multimodal computer games involving that subject's content. She provided "10 design principles for implementation 'The Globaloria Way.'" For example, developing educational games requires students to spend significant time, engaging daily on personally chosen projects involving open-ended and creative design tasks. A transparent and collaborative studio environment facilitates the sharing of work and provides many opportunities for social expression and discussion about game projects. Students thus learn through four modes simultaneously: (1) through design and teaching, (2) through peer-to-peer interactions, (3) through co-learning with teachers (and also from watching the teachers themselves learn), and (4) from online research and consultation with other experts (just-in-time learning) via pre-scheduled web conferencing and a help desk. (See Figure 2.3.)

The basic technology underlying the Globaloria platform is open-source MediaWiki with customized MediaWiki extensions, PHP, MySQL, Tumblr, Blogger and multiple Google tools. Students learn to program their games much like professionals in the real world using Adobe Flash Actionscript. The World Wide Workshop Foundation's team (creators of Globaloria) chose Flash for students' programming for a number of reasons, including:

- They themselves are expert developers in Flash;
- Flash provides a wide variety of tools, such as interfaces and video tutorials, to support users and thus can support a range of skill levels from novice to professional;
- Flash's capability is present on many websites and in simulations and media devices;
- Flash is an industry professional standard in game development and multimedia programming, and so proficiency in Flash is likely to help provide students with internships and job opportunities in the future.

Finally, Caperton described research she and colleagues conducted on the impact of implementing models of Globaloria for fostering computational thinking and inventiveness among low-income rural students and low-income minority urban schoolchildren: (1) Model 1 in 45 schools throughout the public school system in 20 counties in the state of West

Virginia, where 1,300 students in rural middle schools, high schools, community colleges, and alternative education institutions participated with 55 educators in 2010 for credit and a grade; and (2) Model 2 within a charter middle school system in East Austin, Texas, where every student in that school took Globaloria once a day for 90 minutes for the entire school year. She provided an overview of selected research results[7] and shared video case studies.[8] Caperton argued that these were powerful demonstrations of plausible paths and activities for teaching computational thinking concepts to low-income rural and urban students of underserved communities.

4.2 COMPUTATIONAL THINKING AND TECHNOLOGY

4.2.1 Questions Addressed

- What are the relevant lessons learned and best practices for improving computational thinking in K-12 education?
- What are examples of computational thinking and how, if at all, does computational thinking vary by discipline at the K-12 level?
- What exposures and experiences contribute to developing computational thinking in the disciplines?
- How do computers and programming fit into computational thinking?
- What are plausible paths and activities for teaching the most important computational thinking concepts?

Presenters:
   Robert Panoff, Shodor Education Foundation
   Stephen Uzzo, New York Hall of Science
   Jill Denner, Education, Training, Research Associates

Committee respondent: *Yasmin Kafai*

4.2.2 Robert Panoff, Shodor Education Foundation

Robert M. Panoff, founder and executive director of the Shodor Education Foundation, is a proponent of teaching computational thinking through computational science. At the same time, he stresses the

---

[7] For more information see www.WorldWideWorkshop.org/reports. Last accessed February 7, 2011.
[8] For more information see www.worldwideworkshop.org/programs/globaloria/vftf. Last accessed February 7, 2011.

importance of certain metacognitive skills—in particular, being able to know that something learned (e.g., through computation) is right. Panoff described quantitative reasoning and multiscale modeling as components of computational thinking

Quantitative reasoning is not necessarily computer-related, but it is essential for anyone to make sense out of using a computer. An impediment to quantitative reasoning noted by Panoff is that many individuals have inconsistent and faulty intuition about quantity. For example, he pointed out that many people believe that two-fifths (2/5) is a small number, whereas 40 percent feels like a large number to them. He said that one metropolitan police department assigned more officers to patrols on Friday and Saturday night because a careful analysis of the data had shown that just under 30 percent of the car break-ins were on either a Friday or a Saturday night. But since 2/7 is 29 percent, the frequency of car break-ins was actually consistent across weekdays and weekends. (Panoff further noted that engaging in computational thinking is a partial remedy to misconceptions about quantity.)

Panoff described an exploration based on quantitative reasoning that addressed computational thinking and algorithmic thinking. Consider the number given by 355/113, and then explore the algebraic identity given by 355/113 − 101/113 − 101/113 − 101/113 − 52/113. In principle, this quantity should equal zero. But it does not when evaluated on a calculator. Panoff noted that most students realize that "something's not right" when they are confronted with this "identity," and he maintained that such a realization is the beginning of a serious exploration of how numbers are represented in a computer.

A second example involved calculators. A person who types the expression "3 + 2 × 6" into Google will obtain the answer 15, whereas the same expression typed into some calculators (such as the Accessories calculator of Windows) will yield the answer 30. Understanding why such a difference exists is challenging to some students. Another illustration is calculating the sum of A/B + A/C + A/D + A/E on a calculator. A student can perform each of these operations individually, or she can factor out A to obtain A × (1/B + 1/C + 1/D + 1/E). Again, these two sums are identical algebraically, but the algorithm (i.e., the specific steps to be taken in a particular sequence) is different and simpler in the second case than in the first.

Panoff's third example requires understanding of orders of magnitude. He illustrated the point by asking what a student needs to know in order to answer the question "How much bigger is Earth than Pluto?" An obvious way to approach this problem is to perform Internet searches for the mass of Earth and the mass of Pluto. But an Internet search for the mass of Earth generates 20 or 30 different values, which have a spread of several percent. How does one know which value to use?

Here context matters—why is one asking the question about relative sizes? If the question relates to how big an object has to be in order to be a planet, then in the absence of a formal definition of planet, one only needs to know that the ratio M_{Earth}/M_{Pluto} is on the order of a few hundred—and a difference of "several percent" is simply irrelevant to knowing which value of M_{Earth} to use.

In Panoff's fourth example, he pointed out that many people (in this case, medical residents) do not distinguish between "most of the time" and "more often than anything else." For example, a physician may say to the patient that "most of the time, if kidney cancer comes back, it goes to the lungs first." In fact, kidney cancer goes to the lungs 28 percent of the time, which is more often than anyplace else, but 72 percent of the time (i.e., most of the time), it goes somewhere else.

As for multiscale modeling, Panoff argued that technology enables one to re-present data and relationships (noting that one meaning of representation is to re-present). He illustrated by considering the Lennard-Jones potential function:

$$V(r) = k \times ((S/r)^{1/12} - (S/r)^{1/6}).$$

When $r = S$, the potential V is equal to zero, and so $r = S$ defines the point at which the function crosses the horizontal axis. However, changing the value of the parameter S has two effects on the shape of the curve—the location of the crossing point and also the width of the potential well. The second effect is apparent most easily by graphing the function interactively, varying the value of S.

As for pedagogy, Panoff's programs entail a learning progression of students running models at first, moving to modifying models, and then in a culminating step writing their own models. For example, a student might run a model and then manipulate the model's parameters in order to explore what happens and to make conjectures about what would happen when a parameter is changed. Then she might modify it by moving a slider bar, or two or three slider bars. And then she might change the number of slider bars. Finally, she will write a model that calls for the use of slider bars to change parameters.

Pedagogically useful computational models are accurately implemented and provide appropriate data visualization tools. They are controlled by the student user and are honestly described (i.e., the description includes information about the flaws and limitations of the model), although other students and faculty and the scientific community at large collaborate with unit authors to develop the models. Last, they are coded with the goal that they can be extended by another party (students, in particular). The content provided in the models is based on common texts and national standards.

4.2.3 Stephen Uzzo, New York Hall of Science, Museum Studies

Stephen Uzzo from the New York Hall of Science (NYSCI)[9] talked at the end of the workshop about the transformational effect that data-rich science has in computational thinking and about some ways to better prepare future scientists. He noted that data overload is a central theme of 21st century science. Data are accumulated in enormous quantities for biomedical, environmental, and social science applications, enabled by the rapid growth in computing power and sensing technologies.[10] Highlighting this data overload, Uzzo pointed to various statistics such as that some science disciplines produce more than 40,000 papers a month, and computer users worldwide generate enough digital data every 15 minutes to fill the Library of Congress.[11]

In the face of such overload, Uzzo suggested, the traditional method of science—modeling natural phenomena and then validating those models against data gathered from nature—is inadequate. This traditional method assumes an environment in which data are relatively scarce, whereas much of science today is characterized by data in volume. A new approach, "e-science," is needed in this environment.[12] E-science focuses on managing, modeling, and making discoveries in massive amounts of captured data; seeking patterns; and identifying dynamics, influences, and complex and emergent behavior in whole systems.

Uzzo further argued that the computational thinking needed to engage in e-science includes a number of often-neglected concepts:

- *Complexity*. Practitioners need to know when the e-science paradigm for doing scientific research is (and is not) more appropriate than other paradigms of research (theory, experiment).
- *Data visualization*. Because of the large volumes of data involved in e-science research, visualization (and human interpretation of the resulting images) may be a more effective method for detecting and identi-

---

[9] Hall of Science activities entail developing exhibitions and educational programs for STEM learning, and evaluating them for pedagogical efficacy in conveying the relevant concepts to the public and to K-12 students.

[10] Input technologies such as efficient, small, and cheap sensors; automated logging systems; high-resolution remote sensing from satellites; robotics systems for DNA sequencing; protein mass spectrometry; and functional magnetic resonance imaging (fMRI) are just a few examples.

[11] Manish Parashar, 2009, "Transformation of Science Through Cyberinfrastructure: Keynote Address," presentation at Open Grid Forum, Banff, Alberta, Canada, October 14, 2009. Available at http://www. ogf.org/OGF27/materials/1816/parashar-summit-09.pdf. Last accessed February 7, 2011.

[12] Tony Hey, Stewart Tansley, and Kristin Tolle, eds., 2009, *The Fourth Paradigm: Data Intensive Scientific Discovery*. Redmond, Wash.: Microsoft Research.

fying patterns than are traditional methods of reductive data analysis. Practitioners may need to develop new visual metaphors that are better for revealing patterns in complex data and techniques for displaying and comparing large amounts of data.

- *Network science*. In e-science, theoretical generalizations may be based on network science, the study of properties and behaviors of complex, dynamic systems of interaction. One often sees similar network functions and structures emerge across a variety of different problem domains.
- *Data interoperability, data sharing, and other collaboration skills*. Practitioners of e-science must understand many kinds of shared data types and the technical issues in data sharing and data interoperability that inevitably come up in collaborating with other practitioners and across divergent fields of study.
- *Using semantics for creating more effective data structures*. E-science places a premium on the ability to find general patterns in phenomena and then to identify similar instantiations in examining other phenomena. For such purposes, the use of Boolean logic for combining and parsing large amounts of data is insufficient. Searches based on Boolean logic are also ineffective with large amounts of data because both false positives and false negatives are problematic. Although search engines may work well for fact-finding, they do not serve well to identify patterns, trends, or outliers. Perhaps more importantly, the context in which a piece of knowledge was created or can be used may be missing, making intelligent data selection, prioritization, and quality judgments extremely difficult.

Semantic approaches are needed to deal with data at large scale. (Biomedicine is a canonical example of a domain in which this is true.) A semantic web uses *triples* instead of search terms. A triple consists of two ideas (the first two elements of the triple) that are linked through a term describing how the ideas are related (the third element).

E-science requires a cyberinfrastructure capable of processing data in prodigious quantity and of making large data sets available to researchers reliably and promptly. It must facilitate interoperability between applications used by researchers, and it must provide easy-to-use tools for processing, manipulating, and combining multiple data types. In discussion, Al Aho noted that "the software world of today is largely a Tower of Babel with lots of incompatible infrastructures and a lot of expense regarding who pays, who collects the data, who maintains the data, who maintains and evolves the software."

To illustrate the tools necessary, Uzzo discussed the idea of a "macroscope" and an existing tool called the FreeSpace Manager. The macroscope is an expandable and integrated set of applications that scientists can use to share scientific data sets and algorithms and to assemble them into workflows.[13] Uzzo and NYSCI, in cooperation with the School of Library and Information Science at Indiana University, are developing systems that allow museum visitors to create, format, present, and mine data the way scientists do. The macroscope would help to identify patterns, trends, and outliers in multiple large-scale data sets, whether static or streaming.[14] Such tools can continuously evolve as scientists add and upgrade existing plug-ins and remove obsolete ones—all with little or no help from computer scientists.

To support collaborative data sharing involving multiple data types and streaming, the University of Illinois at Chicago is developing the Scalable Adaptive Graphics Environment (SAGE), a central element of which is the FreeSpace Manager.[15] SAGE is a physical room whose walls are made from seamless ultra-high-resolution displays fed by data streamed over ultra-high-speed networks from distantly located visualization and storage servers. SAGE allows local and distributed groups of researchers to work on large distributed heterogeneous data sets. (To illustrate, users could be simultaneously viewing high-resolution aerial or satellite imagery, as well as volumetric information on earthquakes and groundwater.) The FreeSpace Manager provides an easily understood and intuitive interface for moving and resizing graphics on the display, giving users the illusion that they are working on one continuous computer screen, even though each of their systems is physically separate. The FreeSpace Manager is similar to a traditional desktop manager in a windowing system, except that it can scale from a single tablet PC screen to a desktop spanning more than 100 million pixel displays.

Uzzo noted that he sees increasing demand for using these kinds of sophisticated tools in the Hall of Science, not only for accessing data sets virtually within the museum walls, but also for bringing such tools into remote K-12 science classrooms through NYSCI's Virtual Visit teleconferencing program. In the past, NYSCI outreach efforts were based solely on synchronous interactions between a museum facilitator and classroom students. However, because of the complexity of the science NYSCI is

---

[13] Joël De Rosnay, 1975, *The Macroscope*, New York: Harper & Row Publishers.

[14] Katy Börner, 2011, "Plug-and-Play Macroscopes," *Communications of the ACM* 54(3):60-69. Available at http://ivl.slis.indiana.edu/km/pub/2010-borner-macroscopes-cacm.pdf. Last accessed February 7, 2011.

[15] Andrew Johnson, Jason Leigh, Luc Renambot, Arun Rao, Rajvikram Singh, Byungil Jeong, Naveen Krishnaprasad, Venkatram Vishwanath, Vaidya Chandrasekhar, Nicholas Schwarz, Allan Spale, Charles Zhang, and Gideon Goldman, 2004, "LambdaVision and SAGE—Harnessing 100 Megapixels," presentation at the CSCW Workshop on Human Factors in Advanced Collaborative Environments, Chicago, November 6, 2004. Available at http://www.evl.uic.edu/aej/papers/CSCW-SAGE.pdf. Last accessed February 7, 2011.

teaching today, many data inputs are needed, and the remotely located students need to be able to share and interact with those data as well.

In general, Uzzo argued for teaching a new generation of science students these kinds of e-science data processing and interaction skills, thereby creating the demand side for the infrastructure that e-science will need to succeed. He further suggested that informal learning institutions may be in the best position to advance the cause of e-science because these institutions have an opportunity to move computational thinking beyond the traditional bounds of today's computer science by helping to close the gap between science as a research activity and learning about science. These institutions are also in a good position to conduct learning research around this topic and then to integrate such research into professional development and curriculum development for K-12 formal education.

4.2.4 Jill Denner, Education, Training, Research Associates

For Jill Denner, a developmental psychologist with Education, Training, Research (ETR) Associates, the programming of computer games provides an appropriate context for the development of computational thinking in middle school students.

Denner and her colleague Linda Werner, a computer scientist from the University of California, Santa Cruz, argue that the programming of computer games connects to computational thinking in several ways. One important connection is in the modeling of abstractions—in Denner's words, "Youth are engaging in modeling abstraction while programming a game when they create a model of their make-believe world, which includes creating variables, new methods, and thinking at multiple levels of abstraction, such as how the player will interact with the game and what the goal of the game is." A second important connection between programming of computer games and computational thinking is to algorithmic thinking. To make their games playable in the way they envision them, they must understand when and how to program using sequential, parallel, or conditional execution, and how to create a logical process through which a player can interact with the game.

In one pilot study involving 30 students using Storytelling Alice to develop 23 different games, Denner reported that students used specific programming constructs showing evidence of computational thinking (i.e., algorithmic thinking, abstraction and modeling) such as event handling, parallelism, additional methods, parameters, alternation, iteration, and conditional execution:

> Many of the students created their own methods and used parameters which we see as examples of modeling and abstraction. In their final games there was limited use of alternation and limited use of iteration.

In part this is due to how we taught them when we bundled the if/else construct with another complex programming construct that made it more complicated for them to learn it. We feel they didn't incorporate loops due to motivation; many of the students didn't see the point of creating loops when they could just repeat code segments. They didn't see the point of creating more efficiency.

Denner and Werner's approach is based on students engaging with computer games along a use-modify-create continuum. First, they play other students' games and work through three tutorials that teach programming with Alice. The goal of the "use" phase is for students to learn about the Alice interface and the kinds of games that they might make. Second, students learn to modify an existing game through a series of graduated self-paced challenges. The goal of the "modify" phase is to experiment with different strategies and their results, and to build an understanding of the mechanisms that they will use to program a game. Last, students create an original game de novo.

Denner reported on several lessons learned from the project:

- *Individual differences matter a great deal*. Denner pointed out that students have different starting levels, willingness to fail, and motivations. Some students prefer to learn by playing around, whereas others prefer to follow step-by-step instructions to carry out a task. Some students are afraid to fail and thus are unwilling to tackle problems that entail the risk of failure (e.g., using a concept incorrectly). Other students are intrepid explorers who are curious, creative, and undaunted if and when they fail at doing something. Those unwilling to explore a range of strategies are unlikely to get beyond modification of an existing program and will thus never create a truly original game. Denner found it necessary to balance student engagement on a problem with motivating them to learn more complex or difficult concepts needed for their programs. Specifically, she suggests that to promote computational thinking during computer game design, teachers must:
  —Be strategic with examples: students use what they see.
  —Provide graduated instructional materials that can accommodate a range of programming experience and styles.
  —Balance structure with exploration. It is important to encourage authentic interest, but also to provide enough structure to encourage games that include computational thinking concepts.
- *Students program differently in pairs than by themselves*. Compared to students working individually, students in pairs spent more time doing programming and housekeeping tasks (e.g., saving and testing their code), whereas the students working by themselves spent more time doing things like screen layout, changing the appearance of the game, and

adding objects. For most students, pair programming is highly motivating and improves their ability to communicate concepts. When students have to work directly with a partner next to them on one computer, they have to explain their complex ideas simply so their partner understands. The quality of pair interaction determines the extent to which the students engage in computational thinking and persist in the face of challenges.

- *The measurement of computational thinking requires multiple sources of information.* Denner and Werner are analyzing several sources, including computer logging data that show what students are doing when programming a game. They also give students performance assessments to measure algorithmic thinking and abstraction, and code student games for frequency of aspects of computational thinking, as well as computational thinking patterns.

Most of this research has focused on groups that are underrepresented in computing—girls and Latinos. Denner reported that they faced a number of challenges in their middle school efforts to promote computational thinking among students in both high- and low-resourced schools. Challenges included mundane issues such as difficulties with hardware and software and with Internet access. Other challenges were how to create effective instructional materials to help teachers with little or no training in computational thinking support it among their students. Finally, they faced the challenge of motivating students to engage in sustained complex thinking in an after-school setting.

4.2.5 Lou Gross, National Institute for Mathematical and Biological Synthesis

Lou Gross directs the National Institute for Mathematical and Biological Synthesis (NIMBioS), an organization supported by the National Science Foundation and by the Departments of Homeland Security and Agriculture.[16] The primary goals of NIMBioS are to foster the maturation of cross-disciplinary approaches in mathematical biology and the development of interdisciplinary researchers who address fundamental and applied biological questions. NIMBioS has an education and outreach program that offers a variety of activities for K-12 students and teachers, university and college students and faculty, professional industry audiences, and the general public. These activities focus on education at the interface of mathematics and biology.

---

[16] More information about and further description of the National Institute for Mathematical and Biological Synthesis (NIMBioS) can be found at "NIMBioS," website, http://www.nimbios.org. Last accessed February 7, 2011.

Gross argued that key to computational thinking is tying a computational worldview to a student's everyday experience. In his words, "Student comprehension of computation and appreciation for its importance in everyday experience would be enhanced at every level of the educational experience if we encourage connections between computation and the models (internal to their experience, as well as those used to understand scientific processes) students use, and the data they collect from their own observations of the world around them."

To support this perspective, he offered several examples.

His first example involved an everyday problem—how to pick a checkout line at a grocery store. What variables might affect one's decision? Workshop participants suggested line length, the presence or absence of a bagger or of someone writing a check, the number of items in a person's cart, and whether the line is an express line. Gross pointed out that high school students often ask about the presence or absence of someone cute in the checkout line, thus illustrating the point that the criteria for decision making depend on the nature of the model involved and its purpose.

His second example involved a simple game about which students are asked to make a prediction. The game involves groups of three to five students with a cup containing one blue and one yellow bead, and a separate supply of blue and yellow beads. Students are asked to pull a bead from the cup at random and then to place it back into the cup along with another bead of the same color, and then to repeat this procedure until there are 20 beads in the cup. Students are asked to predict what will happen in their cup. (An important advantage of this example is that it is easy for students to collect data in a single class.)

A typical prediction is that the cup will become mostly blue or mostly yellow depending on the color of the bead first chosen. What happens in fact is that for any given cup, the percentage of blue beads converges to a fixed fraction, but the fraction is different for different cups. With an enormous number of cups, the result is that nearly every fraction from zero to one is represented, and in the limit of an infinite number of cups, the distribution from zero to one is uniform. As for the typical prediction, the actual result is that it is true that a blue bead being drawn first does result in a greater likelihood that the ultimate fraction of blue beads is high, but this is not true all of the time.

The cup scheme described is known as a Polya's urn, and the resulting sequence of bead configurations is described by a Martingale process—the system has built-in dependencies and a feedback structure that is a very common property of biological systems at many levels. Gross uses this example to extract three primary conclusions:

- *Randomness can lead to order.* Although the underlying system is random (because the color of the beads drawn cannot be predicted exactly), each cup approaches having a fixed fraction of each bead, but the fraction is different in different cups.
- *Randomness can lead to complexity.* By combining outcomes from different types of beads, highly complex outcomes are possible. That is, with a large number of beads and many beads drawn, one can develop any number of stable outcomes.
- *Because random processes can produce outcomes of such complexity, once natural or artificial selection is added, the result is a powerful mechanism to explain the complexity of the world around us.* This is the basis of much of the explanatory power of modern biology.

A third example provided by Gross involved descriptive statistics using personal data. The question posed to students is what happens to one's height overnight. Students make predictions about their change in height and consider factors that might affect height, such as their height when they go to bed, their gender, the amount of sleep they had last night, the amount of alcohol they consumed last night, and so on. They collect these data over four nights and then explore the data using basic descriptive statistics, bar charts, scatter plots, and regression analysis.

The resulting data set is easily understood, is multivariate, has potential multiple causal factors, and illustrates problems with sampling and outlier effects. For example, a height measurement might be recorded as 800 mm. Since everyone realizes that no one in the group is only 800 mm in height, the example illustrates the point that bad data are sometimes collected and must be discarded. Graphing the data in different ways to show possible relationships between different combinations of two variables provides insight into multivariate relationships, thus illustrating that simple descriptors (e.g., measures of central tendency and dispersion) may not be adequate to describe what's going on.

The example can also motivate a discussion of the importance of institutional review boards in approving experiments dealing with human subjects.

The final example was intended to introduce students to mathematical notions of vectors, matrices, Markov chains, equilibrium, and stability. Gross's example began with an aerial image of Washington, D.C., pulled from Google Earth. He asks students, "How would you describe this image?"

The image shows buildings, roads, trees, and other typical topographic features. But students eventually describe the image by saying how much of the image is this color or that color, how much is in buildings, how much is in roads, and so on. They basically then figure out that

that's what a vector is, that they're describing the image as a vector in which the components consist of the fraction of the image that is of each type. One interpretation of this vector is that it represents a probability distribution of the landscape for a discrete number of components. (Gross pointed out that the idea of using a vector to represent a multidimensional entity is independent of any discussion of probability.) Students also realize (with some coaching) that spatial aspects of the image are not included in the vector description. For example, the vector does not change if some of the buildings are moved around in the image. However, a vector description of an image taken in 1988 would differ from one of an image of the "same" scene in 1949. That is, a time-varying vector describing a scene is one way to characterize change in land use over time. Students are then able to derive the basics of matrix multiplication by finding the fraction of each landscape type that transitions to another type in each time period, the transition matrix of the Markov chain, and use this to determine the landscape vector at future times. After a long time, corresponding to many matrix multiplications, the landscape vector is near equilibrium, given by the eigenvector for the dominant eigenvalue (one) of the Markov chain.

With this background in hand, Gross uses with his students prepackaged software to demonstrate computational methods of looking at change across a landscape, e.g., coupling between an image, a dynamically changing vector, in this case a bar graph, and then an overall descriptor. An example of prepackaged software for this application is found in EcoBeaker, a set of computational laboratories useful for analyzing ecological data, and students also use code they develop in Matlab to analyze different types of landscapes.

4.3 COMPUTATIONAL THINKING IN ENGINEERING AND COMPUTER SCIENCE

4.3.1 Questions Addressed

- What are the relevant lessons learned and best practices for improving computational thinking in K-12 education?
- What are examples of computational thinking and how, if at all, does computational thinking vary by discipline at the K-12 level?
- What exposures and experiences contribute to developing computational thinking in the disciplines?
- How do computers and programming fit into computational thinking?
- What are plausible paths and activities for teaching the most important computational thinking concepts?

Presenters:
> Christine Cunningham, Museum of Science, Engineering is Elementary Project
> Taylor Martin, University of Texas at Austin
> Ursula Wolz, College of New Jersey
> Peter Henderson, Butler University

Committee respondent: *Marcia Linn*

4.3.2 Christine Cunningham, Museum of Science, Engineering is Elementary Project

Christine Cunningham from the Museum of Science in Boston spoke about its Engineering is Elementary (EiE) curriculum. The EiE project is developing an elementary school curriculum to help students learn about engineering. It integrates engineering with topics in elementary school science. EiE also conducts professional development of educators. The project has four goals:

- *To increase children's technological literacy.* This is the primary driving idea underlying the project.
- *To increase elementary educators' ability to teach engineering and technology.* The Museum of Science realized early that the first goal required teachers who understood something about engineering and technology, and that very few elementary school teachers have ever had any exposure to formal engineering.
- *To increase the number of schools in the United States that include engineering at the elementary level.* To introduce technology education into schools, it is necessary to convince schools and districts that there is actually room for it in the curriculum as they currently teach it.
- *To conduct research and assessment to advance the first three goals and contribute knowledge about engineering teaching and learning at the elementary level.* Having as well as presenting research and assessment data is necessary to persuade schools—and to the extent that such data can relate to other topics being taught, so much the better.

Cunningham went on to discuss the project's lessons learned and the resulting best practices. The first principle was the importance of listening to teachers and involving them in every aspect of the development process. For example, because teachers are responsible for covering content that is largely prescribed by external influences, any new content (e.g., new disciplines or concepts) must integrate with or reinforce content or topics already being taught. Cunningham and her colleagues identified 20 topics that are commonly covered in elementary science programs, paired each with an engineering field, and illustrated the pairing with a particular technological device or process (Table 4.1). Cunningham stressed that understanding engineering habits of mind and mental processes is an important aspect of their work. They address this goal by developing specific curriculum units that focus on processes.

A second principle is to build on what teachers know or feel comfortable doing. It is well known that many elementary school teachers are

TABLE 4.1 Correspondences Between Elementary Science and Engineering

Topic from Elementary Science Program	Engineering Specialty	Corresponding Technological Device or Process
Water	Environmental	Water filters
Insects and plants	Agricultural	Pollinators
Wind and weather	Mechanical	Windmills
Simple machines	Industrial	Chip factory design
Earth materials	Materials	Walls
Balance and forces	Civil	Bridges
Sound	Acoustical	Sound representation
Organisms	Bioengineering	Model membranes
Electricity	Electrical	Alarm circuits
Solids and liquids	Chemical	Playdough process
Landforms	Geotechnical	Bridge sitting
Plants	Package	Plant package
Magnetism	Transportation	Maglev vehicle
Energy	Sustainable	Solar cooker
Solar system	Aerospace	Parachute
Rocks and minerals	Materials	Replicate an artifact
Floating and sinking	Oceans	Submersible
Ecosystems	Environmental	Oil spill remediation
Human body	Biomedical	Knee brace
Light	Optical	Lighting system

uncomfortable with science, and Cunningham found that engineering (and presumably computational thinking) is even more terrifying. To address this issue, Cunningham and colleagues begin their presentations with exercises in literacy—an illustrated storybook for children. The story has significant engineering content but is presented as a reading exercise, so that at a minimum students will receive a very general introduction to the topic. This storybook also provides context for the engineering activities that the kids will be doing in class. Elementary teachers are generally quite comfortable teaching literacy, and so this is a gentle introduction to the new discipline of engineering.

Cunningham also noted the importance of articulating how new content and skills are responsive to existing educational standards, such as those from the International Technology and Engineering Educators Association (ITEEA) Standards for Technological Literacy, the National Science Education Standards from the National Academy of Sciences, and the math standards from the National Council of Teachers of Mathematics. Such standards could include, for example, core concepts of technology such as systems, processes, feedback, controls, and optimization; the design process as a purposeful method of planning practical solutions to problems; inclusion in the design process of such factors as the desired elements and features of a product or system or the limits that are placed on the design; and the need for troubleshooting.

From time to time, learning about engineering can be motivated in terms of meeting teacher goals that are not necessarily based in educational standards. For example, many elementary school teachers want to find ways to help their students work together in teams. Persuading students to work together, to play nicely, and to communicate what they're doing is something that many teachers want to accomplish at the beginning of each year, and engineering education can often be an important part of such persuasion.

A related point is the importance of student evaluation. Both teachers and students pay much more attention to material when student understanding of such material will be evaluated. The evaluation process need not be a one-to-one correlation (i.e., we teach X and then students are evaluated on their knowledge of X), but if teaching X helps students to understand Y better, and Y is assessed, teachers will be more likely to continue teaching X.

Another lesson learned is the desirability of starting small. Teachers tend to be more willing to invest a couple of class periods to experiment with a new concept than an entire school semester or year. The success of one individual teacher with a particular concept or topic can catalyze others, as his or her students tell their friends about an interesting new experience in class. Other teachers hear of students' positive reaction and often want to try the concept or topic themselves. These efforts build grassroots support within the school for change.

Professional development and solid curricular materials are also important. Because elementary teachers are inexperienced with the subject matter of engineering, teaching materials have to be explicit and clear. For example, because learning objectives drive the specific experiences or exposures embedded in different curricular units, objectives need to be very explicit and specific—children will know X and be able to do Y—rather than high level and abstract. Learning objectives should also be few in number and relatively simple in scope so that a high degree of student mastery is possible. And the materials must provide ways of specifically assessing the scope and extent of student mastery and comprehension.

As for the pedagogical approach taken to the subject material, Cunningham and colleagues have found that hands-on experiences are particularly important for young learners. They have fielded many requests to replace physically manipulative experiences in handling objects with a click-and-drag interface on the computer that students can use to connect objects on the screen. But knowledge about the physical world that teachers take for granted cannot be assumed in students. For example, students don't necessarily know that a fuzzy pompom will pick up pollen better than a smooth marble. In fact, that fact is engineering knowledge, and it's "common sense" only if one has real-world experience with pompoms and marbles.

Also, context matters a great deal to students, especially to girls and underrepresented minorities, who often lack a cultural frame for why they should care about learning about engineering and what it might be used for. The project takes great care to ensure that its challenges are inviting to students who are often underrepresented in STEM (science, technology, engineering, and mathematics); one core way it does this is by illustrating how engineers (and engineering) might help people, societies, or animals. The storybooks are an essential element of context-setting, but it is important to contextualize the entire learning experience and not just the beginning.

Cunningham closed by pointing out that some of the lessons above for introducing engineering into elementary school also applied to middle schools and high schools. Specifically, she underscored the importance of integrating the new material—in this case, engineering—with other things that these schools are already teaching. After that integration is achieved, the new material may become more primary—but emphasizing its importance as a primary focus from the start is a strategy that is not likely to succeed in getting it introduced in the first place.

4.3.3 Taylor Martin, University of Texas at Austin

Taylor Martin of the University of Texas at Austin discussed several themes that she regarded as important for the teaching of computational thinking:

- *Personal empowerment.* In Martin's view, the teaching of computational thinking to students should impart to them a personal sense that they can in fact undertake intellectual tasks that they initially feel they cannot perform successfully. Underlying this sense is the ability to break a complex task into constituent parts, each of which makes some progress toward the goal. She illustrated by pointing to Web searches that yield information on how to do or fix something and suggesting that a Web search may well be the most sensible "first step" in solving a complex problem. Empowerment also implies that the computational thinker has confidence in being able to "talk" with the computer and getting it to do what he or she wants it to do.
- *Motivation and authenticity.* Martin noted the importance of personally relevant tasks for motivating people to undertake the hard work of learning new ways of thinking and acquiring habits of mind. However, she also pointed to the idea that using computers can be fun and motivating in and of itself for many individuals. Many individuals will explore the ins and outs of a computational device just for the fun of discovering what it can do, and pedagogy should take advantage of this phenomenon. She further observed that exploration of such devices is greatly enhanced when they are ubiquitously present—when the devices are not present, she argued, students are not thinking computationally.
- *Habits of mind.* Martin argued that when someone becomes facile with computational thinking, the notion of computer-as-device disappears, and what remains are the worldview and habits of mind associated with computational thinking. She noted that an experienced computational thinker cannot resist thinking of ways to save effort when repeated actions are required to accomplish a task—they are driven to develop computationally informed approaches to solve these problems. For example, these individuals understand that in large stores, even if there are four cashiers, having one line makes more sense than having four lines.

Regarding the infrastructure needed for teaching computational thinking, Martin said that present trends point to the disappearance of the computer "as computer" in the future—the computer will become increasingly invisible. If so, teachers of computational thinking will have to find pedagogical approaches that do not necessarily depend on the computer per se.

Nevertheless, today computers are a valuable instructional tool when teachers are comfortable with them, when activities are student-centered, and when enough equipment is available. From her perspective, schools do have many computers—but these computers are not well matched to the pedagogical tasks at hand. Martin is an advocate of an infrastructure based on open-source software, and she believes that it makes sense to install such computers in classrooms and see what students and teachers do with them. She also offered an important qualifier—if the resulting computational problem-solving environment does not have the tools that students would choose to use, e.g., Facebook, Gmail, and so on, and the unavailability of familiar tools is likely to inhibit student learning.

With such an infrastructure, Martin's goal is to make the underlying technology as transparent as possible to students, and thus "computational thinking" can be sneaked into student activities without intimidating them so that the computer is "a tool like a pencil, no big deal at all, an extension of your hands."

4.3.4 Ursula Wolz, College of New Jersey

Ursula Wolz described the use of journalism education and the language arts as vehicles for exploring computational thinking in a program at the Fisher Middle School in Ewing, New Jersey.[17] Paraphrasing Gerald Sussman's statements at the first NRC Workshop on Computational Thinking,[18] she began her presentation by arguing that computational thinking requires first and foremost a language through which to express that thinking. Languages can be natural or formal. Language arts instruction focuses on the former, whereas mathematics instruction focuses on the latter. The emphasis on tying programming to mathematics instruction—adapting pedagogical strategies, curricular organization, and assessment methods from math—may lead math-averse students to believe that they can't think computationally.

Wolz argued that demonstrating the relationship between computational thinking and language arts can facilitate integration of computational thinking into the mainstream curriculum. Essential to this enterprise is acknowledgment that computing must become as ubiquitous and integrated as the life sciences, starting with a computational analog to the butterfly chrysalis in preschool. But computing must be infused into all

---

[17] For more information see the Interactive Journalism Institute, website, College of New Jersey, www.tcnj.edu/~ijims/ or http://www.bpcportal.org/. Last accessed February 7, 2011.

[18] NRC, 2010, *Report of a Workshop on the Scope and Nature of Computational Thinking*, Washington, D.C.: The National Academies Press. Available at http://www.nap.edu/catalog.php?record_id=12840. Last accessed February 7, 2011.

curricular areas so that it is not compartmentalized into a "special" activity such as art or gym. Its ubiquity should be through creative expression rather than curricular "chunking" of disassociated content. She noted the problem that adding computing to an overburdened curriculum requires taking something out (often the arts).

Teacher-initiated curriculum development confirms Wolz's contention that learning programming and computational thinking must be contextualized. Scholars of computer science may study and examine formal languages in the abstract, leading to the traditional focus in programming courses on the constructs of a language (e.g., variables, expressions, loops, functions). The analog in modern language instruction would be to require elementary students to diagram sentences and master English grammar before being allowed to read literature or write stories of their own. Integrating computational thinking into the language arts curriculum affords students a natural arena in which to practice reading and writing in a formal language (e.g., Scratch) in a meaningful and motivating context.

Wolz argued that language arts programs are inherently flexible, thus inviting innovation. Journalism provides an ideal venue for civic engagement and what Seymour Papert called "serious fun."[19] Language arts is secure in K-12 curricula, and so hitching the computational thinking wagon to language arts helps to ensure that there will be a place for it in an increasingly packed curriculum. Further, 21st century literacy will require facility with as yet unimagined modes of expression that involve computational thinking.

The extracurricular program her project developed reinforces language arts skills and computational thinking by providing a collaborative model for a 21st century newsroom. Teacher-editors and student reporters are assigned a "beat" (e.g., politics, sports, business), research and develop a story, and then create text, graphics, video, and procedural animations in Scratch to post newsworthy stories on the Internet.

Journalism provides an operational context—that of principled storytelling and information dissemination in which students, as constructors of aggregated content (rather than just consumers), must inquire, create, build, invent, polish, and publish. All of these same notions arise in computational thinking as well. Wolz's students iterate on defining the problem, researching it, drafting a solution, and testing it (in the language of journalism, they research, interview, draft, copy edit, and fact check). In the end, they publish.

---

[19] Seymour Papert, "Hard Fun," available at http://www.papert.org/articles/HardFun.html. Last accessed February 7, 2011.

For Wolz, the correspondence between journalism and computational thinking is deeper than mere process. Both are concerned with information access, aggregation, and synthesis (e.g., fact gathering, analysis). In both domains, there are huge concerns about reliability, privacy, and accuracy. Algorithm design is the computational thinking analog for the need for logical consistency in journalism. (Increasingly news articles have a quantitative aspect to them and are expected to contain logically consistent arguments based on reliable data.) Knowledge representation and the appropriate granularity are important as well. Last, she noted the importance of abstraction from cases in both domains.

The middle school teachers with whom Wolz collaborates took the initiative to bring the Interactive Journalism Institute into their classrooms. They view programming in Scratch as a mode of expression through which students practice language arts skills. The teachers integrate computational thinking concepts via Scratch projects and kinesthetic exposure to algorithms (à la CS Unplugged[20]) into curricula ranging from formal report writing to poetry. In technology and math classes, Scratch storytelling projects are used as rewards for completing assignments. Because of the gentle process through which Scratch programming and computational thinking are infused, the math-averse are not dissuaded, and those who want to improve their writing skills are encouraged. Crucial to this approach is minimizing informal coaching and emphasizing student-initiated project selection over didactic instruction in computational thinking. Enthusiasm for this approach is measured in the cultural diffusion throughout the school. Although only 6 teachers and approximately 50 students participated in the extracurricular program, 12 teachers included computational thinking in classes that reach approximately half of the school population of 900.[21]

Wolz believes her program demonstrates that reading teachers also have the capacity to teach elements of computer science effectively with the right support and tools. Three open questions remain: (1) How can computational thinking skills be assessed within this context? (2) How can the impact of computational thinking on language arts skills be assessed? and (3) How can this pedagogy be applied to other primary disciplines such as social studies, science, and math?

---

[20] CS Unplugged, website, http://csunplugged.com/. Last accessed February 7, 2011.

[21] Ursula Wolz, Kim Pearson, Monisha Pulimood, Meredith Stone, and Mary Switzer, 2011, "Computational Thinking and Expository Writing in the Middle School: A Novel Approach to Broadening Participation in Computing," *ACM Transactions on Computing Education*, forthcoming.

4.3.5 Peter Henderson, Butler University

Workshop participants extended the discussion started at the first workshop concerning the nature of computational thinking and computational thinkers. In one perspective, Peter Henderson, formerly chair of the Department of Computer Science and Software Engineering at Butler University, framed his comments about computational thinking against the background of negative perceptions about computing and computer science. He pointed out the poor quality of software, noting that on a day-to-day level, we have all experienced the frustrations and difficulties of dealing with software artifacts that sometimes work and sometimes don't, unexplained and unexpected system crashes, and so on.

He also noted that much of the news regarding careers in the information technology field is negative. Many parents remember the dot-com boom and bust, and even today, the news is filled with stories about greater outsourcing and offshoring of information technology employment. Rapid changes in technology make it difficult for people working with information technology to keep up their skills. Majoring in information technology is confusing, with all of the options in information systems, management information systems, computer science, informatics, computer engineering, software engineering, and so on.

Finally, Henderson argued that computer science is misunderstood by nearly everyone. He pointed to the NRC report on the first workshop on computational thinking and the lack of consensus on what it is, and suggested that it is no different asking any computer scientist—ask 10 different computer scientists what computer science is and you'll get multiple different answers. So, he asked, "How can we advance the cause of a discipline that we don't understand?"

As a unifying theme, Henderson described computational thinking as generalized problem solving with constraints. He argued that almost every problem-solving activity involves computation of some kind. Citing a Fred Brooks article, "The Computer Scientist as Toolsmith II,"[22] Henderson said that for him, the toolsmith metaphor is a convenient umbrella under which the elements of computer science can be combined and then presented to the public in a manageable way. Computational thinking serves much the same purpose.

Henderson noted that humans learn through the use of concrete examples and through pattern recognition. Specifically, he argued that human understanding begins with concrete examples. Humans then identify patterns common to those examples (i.e., they generalize), they

---

[22] Frederick Brooks, 1996, "The Computer Scientist as Toolsmith II," *Communications of the ACM* 39(3):61-68.

specify those patterns clearly, they verify that those patterns are indeed valid, and then they proceed to name the patterns.

Henderson presented two examples. He described a problem presented on a TV series for preschool students entitled Thomas the Tank Engine. In one situation, Thomas is pulling two cars, one red and one green. They are on a track with a siding (connected on both sides), and the problem is to reverse the order of the two cars. Solving this problem requires the students to develop a computational algorithm.

His second example began with a set of number series: 0, 1; 0, 1, 1; 0, 1, 1, 2; 0, 1, 1, 2, 3; and 0, 1, 1, 2, 3, 5. The pattern is that the next number in the sequence is the sum of the previous two numbers; that is, $N_i = N_{i-1} + N_{i-2}$ with $N_0 = 0$ and $N_1 = 1$. And then we name the sequence—Fibonacci numbers.

Henderson went on to describe computational thinking as generalized problem solving with constraints. That is, almost every problem-solving activity involves computation of some kind. He further noted that discrete mathematics and logic are rich sources of examples and material for computational thinking, and thus that discrete mathematics and logic are the foundational mathematics for computational thinking, useful for reasoning about computational processes.

Thus, Henderson would start with computational thinking activities in pre-K (e.g., reversing the two cars), although for the first several years, the term "computational thinking" would not be introduced explicitly. Only later would the notion of computational thinking be explored as such; this philosophy is consistent with Bertrand Meyer's "successive opening of black boxes" view of learning object-oriented programming.[23] For this preparation, traditional mathematics, discrete mathematics, and logical reasoning must be taught at all levels, and so, for example, an advanced placement course in discrete mathematics would replace the current AP course in computer science. A freshman discrete mathematics sequence would be introduced, similar to that currently present for calculus. This view is consistent with the traditional engineering educational model, which emphasizes the science and math foundations of the discipline early (e.g., physics, chemistry, calculus).

---

[23] Bertrand Meyer, 1993, "Towards an Object-Oriented Curriculum," *Journal of Object-Oriented Programming* 6(2):76-81.

4.4 TEACHING AND LEARNING COMPUTATIONAL THINKING

4.4.1 Questions Addressed

- What is the role of computational thinking in formal and informal educational contexts of K-12 education?
- What are some innovative environments for teaching computational thinking?
- Is there a progression of computational thinking concepts in K-12 education? What are criteria by which to order such a progression? What is the appropriate progression?
- What are plausible paths to teaching the most important computational thinking concepts?
- How do cognitive learning theory and education theory guide the design of instruction intended to foster computational thinking?

Presenters:
Deanna Kuhn, Columbia University
Matthew Stone, Rutgers University
Jim Slotta, University of Toronto
Joyce Malyn-Smith, Education Development Center, Inc.

Committee respondent: *Al Aho*

4.4.2 Deanna Kuhn, Columbia University

Deanna Kuhn, a developmental psychologist at Columbia University, sees young learners as evolving through a number of different intellectual stages regarding scientific thinking, and she discussed how these individuals use data and evidence and how such use is relevant to their facility with scientific thinking.

The first level of development involves accepting the possibility of false belief, a level at which children come to understand that knowledge is constructed by human minds and therefore could be false. Often, children are able to realize the potential for false belief by distinguishing data and evidence from theories and claims. This requires a child to conceive of data as possibly not representing the complete reality—if the child sees data as representing a complete reality, there is nothing to distinguish the data from the claim and thus no task of intellectual coordination.

To illustrate this point, Kuhn described several data-gathering and reasoning exercises for kindergarteners. According to Kuhn, kindergarteners can do simple data representation and analysis. For example, when a class of kindergartners is asked the question, What is your favorite TV show?, students are able to understand the process of collecting the data and developing a bar chart based on the answers. They can also make simple inferences such as "more students like show X than show Y." They are also very good at recognizing simple covariation in causal models, that is, "Did A cause O?" in simple cases.

On the other hand, they have difficulty with covariation in a multivariate context (e.g., Which A caused O?) or a negative antecedent and outcome (e.g., Not A and O). For example, when children were given information about a simple pattern of eating green food or red food and then were asked if there was a relationship between the food eaten and healthy teeth, students could see and recognize the pattern when there was one (find covariation) but could not recognize when there was no pattern (non-covariation). Instead they claimed that there was a pattern. Kuhn added that young children also have trouble making the connection between concepts like non-covariation and non-causality. Kuhn says the students were doing isolated data interpretation of isolated instances and not looking for big patterns.

Kuhn also pointed out that although young children can do fundamental experimental design,[24] they often close their inquiry prematurely. For example, if young learners are asked, "If we want to find out if the mouse that's eating my cheese is big or small, which trap should we use?" and then are offered the options of a small-door trap or a big-door trap, the students can often understand that the small trap is going to be informative whereas the large door is not. But, if they are asked, "If we use a big door on the mouse trap, can we say whether the mouse is big or small?," they tend to say, "Yes, that means the mouse must be big."

Premature closure also sometimes occurs when children are presented with confirming evidence. Children often stop the inquiry at this point, not realizing that the inquiry remains unfinished and that confirming evidence is not sufficient to rule out competing hypotheses.[25] Mitch Resnick has made a similar argument that not just kids but also adults are one-cause thinkers—that even adults identify one cause (of potentially many) and assume a partial inquiry is completely explicative.

Young children can eliminate variables based on simple data patterns. For example, Kuhn noted the work of Alison Gopnik, which suggests that children as young as 3 or 4, when presented with an experimental setting in which monkeys sneeze depending on the presence or absence of different types of flowers, can see the covariation between the blue flowers and

---

[24] Beate Sodian, Deborah Zaitchik, and Susan Carey, 1991, "Young Children's Differentiation of Hypothetical Beliefs from Evidence," *Child Development* 62(4):753-766.

[25] Anne L. Fay and David Klahr, 1996, "Knowing About Guessing and Guessing About Knowing: Preschoolers' Understanding of Indeterminacy," *Child Development* 67:689-716. See works of Klahr and Fay, among others, on problems in premature closure or dealing with indeterminacy in young children.

the sneezing.[26] They can also recognize that the red flower or the yellow flower made no difference. In other words, they can eliminate variables.

Despite the fact that even young children can successfully employ some of the intellectual skills of scientific thinking, they can have a hard time articulating how they know something. In particular, they do not understand the epistemological difference between claim and evidence.[27] Although a claim and data may tell a consistent story, they are not the same thing. For example, while looking at a photo of a boy standing on an award podium with a sign labeled with the number "1" and holding a trophy, a child is asked, "How do you know that this boy won the race?" A child will often answer not with evidence of how he knows (e.g., "He is holding a trophy" or "The podium has a number 1 on it") but with a theory of why the outcome makes sense (e.g., "His sneakers were fast").

Even children up to 12 years old tend to focus on evidence and data fragments that support their story, while ignoring or minimizing those that do not. For example, in explaining what causes an avalanche, a student may report that in case A, it was the slope angle that caused an avalanche. Yet the same student will claim that in case B, the slope angle did not make a difference because the slope angle was small and something else caused the avalanche—that is, these older students are having trouble with distinguishing between a variable and a variable's magnitude. The educational challenge at this level is to help the child see the data as evidence rather than an example of a favored claim. Kuhn argued that when a child exercises control, a sort of meta-awareness over this sorting and attribution process, true scientific thinking can begin.

Finally, Kuhn argued that students' facility with experimental design or controlled comparison develops with engagement and practice, even in the absence of direct instruction (although the environment itself must afford opportunities for practice). That is, such skills are naturally useful in an appropriately rich environment, and students begin to recognize and apply these distinctions naturally even without a lot of explicit instruction.

4.4.3 Matthew Stone, Rutgers University

Matthew Stone is a computational linguist at the Rutgers University's Department of Computer Science and Center for Cognitive Science. Stone works at the undergraduate level developing light programming courses

---

[26] Alison Gopnik and Laura Schulz, 2004, "Mechanisms of Theory Formation in Young Children," *Trends in Cognitive Sciences* 8(8):1364-1366.

[27] Deanna Kuhn and Susan Pearsall, 2000, "Developmental Origins of Scientific Thinking," *Journal of Cognition and Development* 1:113-129.

geared to non-computer science majors, particularly in the humanities and social sciences, and to those who tried to avoid math in high school. His efforts explore alternatives to programming-focused curricula to teach these students computational thinking concepts.

Stone emphasized the importance of three key ideas in teaching computational thinking:

- *The universality of computing devices.* Universality explains why it can ultimately be easier to design a machine that does many things (or everything) rather than one that just does one particular thing. Universality also underlies important concepts in computer science such as Turing's theorem, as well as why and how programming languages are useful. In a class oriented toward non-computer science majors, it is impractical and overwhelming to discuss building digital logic, but Jacquard looms to control the weaving of patterns are well within their reach.
- *Algorithmic approaches to problem solving.* The notion of an algorithm as a deterministic way of problem solving is of course important, but an algorithmic approach to problem solving calls for fitting different algorithms together in an overall solution in a way that is worth the effort of doing so. This approach to problem solving explains why people will pay for programs and pay for programmers to write them. It provides historical context for the origins and evolution in society of computing, sorting, and tabulating. It inspires the right kind of people to think about algorithms for everyday problems, such as choosing which checkout line to go to in the grocery store. Stone reported that non-computer science students found it relatively easy to understand Radix sorts and binary searches. Such examples can be used as a basis for motivating a solution to a problem such as searching through a billion randomized slips to find 60 specific items.
- *The importance of representations as correspondences between symbols and the physical world.* With symbols, one can build mechanistic operations to track truth in the physical world, so that mechanical operations have broader meaning. This point explains why computers can be used for entertainment, music, and video; why biological systems are thought of as carrying information; and why computers appeal to cognitive scientists as a model of the mind. It is likely too difficult to delve in depth into representation at the level of understanding the meaningfulness of symbols in John Searle's Chinese room,[28] but there are many examples of digital representation in everyday life that connect with issues of general-

---

[28] John Searle, 1980, "Minds, Brains and Programs," *Behavioral and Brain Sciences* 3(3):417-457.

ity and algorithms in an information economy. Virtually everyone knows Facebook and Google, and they all know about online banking.

Stone argued that these three ideas are at the core of computational thinking, and they are useful for people who are not programmers but rather engage in work that does not require a mathematical or design background. By exploring these ideas in an elementary non-technical context, Stone felt that he was laying the foundation for allowing an initial understanding to grow into a fuller one.

4.4.4 Jim Slotta, University of Toronto, Ontario Institute for Studies in Education

Jim Slotta's presentation addressed technology environments for K-12 classrooms and how these environments could support different pedagogical models.

Slotta first discussed the Web-based Inquiry Science Environment (WISE), which is intended to provide support (i.e., scaffolding) for inquiry activities in science classrooms. Students in these classrooms work collaboratively on projects that range in duration from 2 days to 4 weeks. A typical WISE project might engage students in designing solutions to problems (e.g., design a desert house that stays warm at night and cool during the day), debating contemporary science controversies (e.g., the causes of declining amphibian populations), or critiquing scientific claims found on websites (e.g., arguments for life on Mars).[29]

Tools and interactive materials provided in the WISE environment support collaborative activities. These tools include "inquiry maps" that provide the student with options for what to do next (e.g., to display a Web page that can be used in support of student designs or debates; to view a WISE notes window or a whiteboard, an online discussion, or journals; or to run an applet for data visualization, a simulation, or a causal map). The WISE environment also includes cognitive guidance to promote reflection and critique. WISE provides embedded assessments of student conceptual understanding of the inquiry processes they use, and it support teachers in adopting pedagogical practices that facilitate inquiry approaches to science education.

Slotta described WISE as a largely successful educational innovation for inquiry that was adopted by tens of thousands of students and teachers. In addition, it enabled research on pedagogical models and patterns,

---

[29] Jim Slotta, 2004,*The Web-based Inquiry Science Environment (WISE): Scaffolding Knowledge Integration in the Science Classroom*, University of California, Berkeley, available at http://tccl.rit.albany.edu/knilt/images/3/36/Slotta_WISE.pdf. Last accessed February 7, 2011.

pioneered authorware and portal technologies, created an "open" curriculum library, and fostered disciplinary partnerships with NASA, the American Physical Society, and the Concord Consortium. Nevertheless, it had a number of significant limitations:

- *Content was not portable across platforms.* Although the curriculum library was open in the sense of being accessible to anyone, anyone who wanted to use elements of it was required to access the content in WISE. Thus, users found it very difficult to make changes to any curricular element.
- *Individual students were unable to interact with their peers in real time in WISE*, which ran within the browser but was unable to interact with other applications that were running on the machine.
- *Implementing contingent behavior in a curricular element was made unnecessarily difficult by the technology.* It is often desirable for an educational application to support execution paths that differ depending on what a student does, but WISE made it hard to design applications to do so.

To address some of these limitations, Slotta and his team engaged with the computer science department to develop a new open-source architecture called SAIL (Scalable Architecture for Interactive Learning) for content display and manipulation that separated the various layers of the learning environment (and in particular separated the content and the user interface) wherever possible.

Using this architecture, WISE was reimplemented. Renamed WISE 3, the reimplementation provided all of the functionality of the original environment but also supported easy interactions with other software, such as the Concord Consortium's Molecular Workbench. On the other hand, it was implemented in Java, which the developers found too limiting from a performance standpoint. The next version, WISE 4, runs on the Web. It retains most SAIL elements, such as portals for managing user groups and the XML structures, as well as some of the metadata, and adds a new presentation layer.

SAIL has been used in a number of other science education efforts as well. For example, SAIL is an integral element of the Science Created by You (SCY) project of the European Union.[30] SCY is a large project that provides a flexible, open-ended learning environment for adolescents. Within this environment—called SCY Lab—students engage in personally meaningful learning activities that can be completed through constructive

---

[30] This discussion of Science Created by You (SCY) includes material found at the SCY website, "Science Created by You," http://www.scy-net.eu/. Last accessed February 7, 2011.

and productive learning. Examples of learning activities include browsing for information, generating a hypothesis, and distributing tasks.

Central to the SCY environment is an emerging learning object (ELO), which is essentially any student-created content that contributes to the learning process. Such content would include notes, reports, simulations, graphs, concept maps, research questions, data, and so on. ELOs are intended to represent the knowledge of a student as he or she learns.

Learning activities are themselves clustered in learning activity spaces (LASs). For example, a LAS titled "Experiment" clusters activities such as "design an experimental procedure," "run experiment," and "interpret data." A LAS also indicates the relationships between learning activities and ELOs.

SCY provides a variety of modeling and simulation tools that support collaborative learning activities. Tools help students producing ELOs and thus determine the type and format of the ELOs, although the student adds the content on his or her own.

Slotta also noted that although the notion of establishing a knowledge community is not new, it has been difficult to implement. The basic idea of a knowledge community is that of students working collectively to aggregate and edit materials in ways that they drive their own learning. It is more open-ended than a traditional classroom, and the community emphasizes patterns of discourse and distributed rather than centralized expertise. But in a curricular environment such as secondary school science (chemistry, physics, for example) that requires coverage of a particular body of subject matter, orchestrating the proceedings in a knowledge community and connecting them to specific learning objectives presents extraordinary challenges.

This complex orchestration of people, materials, resources, groups, conditions, and so on requires a sophisticated technology framework to support it. Slotta developed such a framework, called SAIL SmartSpace (S3).[31] This framework can be regarded as a "smart classroom" infrastructure that facilitates cooperative learning in a milieu of physical and semantic spaces.

From a technical standpoint, S3 supports aggregating, filtering, and representing information on various devices and displays (e.g., handheld devices, laptop computers); locational dependencies (i.e., allowing different things to happen depending on the physical location of a student); interactive learning objects; and an intelligent agent framework. The S3

---

[31] More discussion of S3 can be found in Jim Slotta, undated, *A Technology Framework for Smart Classrooms: Enabling Complex Pedagogical Scripts*, Ontario Institute for Studies in Education, University of Toronto, available at www.stellarnet.eu/index.php/download_file/-/view/558. Last accessed February 7, 2011.

environment is highly customizable and supports the coordination of people, activities, and materials with real-time sensitivity to inputs from students.

As an example of the learning space that S3 can support, Slotta described an S3 environment tailored to a mathematics unit intended to help students understand the relationship between different aspects of mathematics—content categories such as functions, relations, graphing, algebra, and trigonometry. In this implementation, Slotta and his colleagues developed a touch wall at the front of the room where students could interact with these materials. Students worked in groups at their local machines and then the aggregate of their local work appeared both on their local machines and on the touch wall where students could walk up and take turns touching and exploring the space. This environment gave students the ability to manipulate content from different categories and to see relationships between them.

During discussion after his presentation, Slotta noted that one of the advantages of these technology-rich learning environments is that they reduce the intellectual need to consider the technology directly. That is, these environments focus student attention on inquiry, reflection, and collaboration around subject-matter content, rather than on how to interact with the technologies per se—the technology thus becomes more transparent and more invisible to the student.

4.4.5 Joyce Malyn-Smith, Education Development Center, Inc., ITEST Learning Resource Center

Joyce Malyn-Smith from the Education Development Center, Inc., began by noting the importance of designing and managing both school-based and informal learning environments. For learning to occur, she maintained, it is necessary to invite youth into our learning environments and to create a learning exchange.

She suggested that educating K-12 students is different from educating college students. The former have minimal career direction and few internally determined learning goals and objectives. Furthermore, college/university education may be about teaching, but in K-12 environments, successful education is about learning—and in particular, about understanding the learning needs of each and every child in a classroom.

Facilitating learning among today's digital natives is challenging. Malyn-Smith argued that these individuals use their ubiquitous information technologies to learn anything they want to know at any time and anywhere and to any depth and sophistication that their interests and needs take them. Furthermore, their technologies are individualized and mobile, and so they live continuously inside their own portable learning environments. Teachers can exploit the pedagogical opportunities thereby offered by creating a learning exchange during which students and teachers share what they know with each other, especially in non-school contexts.

Citing a white paper titled "Computational Thinking for Youth,"[32] Malyn-Smith said that today's youth often have a substantial familiarity with technology tools and deep understanding of technology concepts that can be foundational to developing their ability to think and solve problems. In addition, one of the main conclusions of that paper is that learners of computational thinking need opportunities for thoughtful, reflective engagement with the phenomena represented.

For example, although nearly every middle school student learns from the textbook that trees help mitigate pollution, students in an after-school program can have a chance to go further, using modeling tools to map the trees in their school yard and record relevant data on species, health, growing conditions, and the like. With this abstraction of their school yard created in the form of maps and data tables, they can use automated models to calculate the benefits of the trees in terms of pollution removal and runoff mitigation. They can also model alternative growth scenarios as they either "plant" new trees, let the existing trees continue to grow, or remove the trees for expanded parking. Re-running the model leverages the power of automation to quickly adjust the underlying parameters and enable seeing what the impacts are. But this iterative process just doesn't fit in a school curriculum packed with hundreds of discrete topics that are connected loosely at best. Time allocations that allow for depth and complexity are part of the culture change needed for computational thinking to take root.

Other advantages of non-school environments include curricular flexibility, staff capacity, and access to infrastructure and to programs, especially in rural areas.

Interrelated challenges have constrained many previous educational innovations, and computational thinking is no different, in Malyn-Smith's view. Addressing any one of these by itself will mitigate limitations, but a successful implementation will require addressing them all.

According to Malyn-Smith, one important consequence of this rich milieu is that today's youth are evolving their own definition for computational thinking through experience. These individuals may not know what to call it, or associate all of the technical terms—but they do know that they are engaged in a way of thinking that is different from that of

---

[32] ITEST Small Group on Computational Thinking, 2010, "Computational Thinking for Youth," Newton, Mass.: Education Development Center, available at http://itestlrc.edc.org/resources/computational-thinking-youth-white-paper. Last accessed February 7, 2011.

people who are not intensive users of technology, and they are applying this way of thinking to the world around them.

Two key research questions arise from this sort of student engagement with computational thinking. First, to what degree and in what ways does the technology expertise of youth contribute to their computational thinking? A related second question is, How and to what degree can the use of technological tools and systems and processes facilitate transfer of learning in STEM careers and sciences?

To understand better what skills and knowledge youth bring to the classroom experience, Malyn-Smith encourages teachers to think broadly about the knowledge base that students are developing in all of their activities, not just those provided in program settings. Although teachers need to know a student's standing relative to the curriculum being taught each year in schools, teachers also should engage in conversations with students about their interests and what they are learning in other settings, such as in museums, through television and radio, by playing games, and through what they're doing with their friends.

Teachers thus play a crucial role in helping students validate what they learn both in and out of school and connect their learning to the standards and benchmarks that define achievement in today's society. Teachers also play a critical role in providing context so that students see the importance of what they learn and how to connect what they know and can do to the skills and knowledge that are valued in society.

As for content, Malyn-Smith argued the need for clarity regarding what computational thinking is about. In the absence of such clarity, "it will be impossible to get any consistency in schools because people won't understand what the topic is about, or people will interpret its definition as seen through only their individual lens." She added that effective nationwide teaching of computational thinking requires a strategic approach based on clear definitions and illustrations rather than a scattershot set of examples.

Malyn-Smith recognized the difficulties in achieving clarity when multiple parties have different views of the essential content. To address these difficulties, she thought that the computational thinking community would benefit from a consensus process to explicate what she called a learning occupation. A learning occupation does not correspond to a specific occupational title or description, but it represents instead the combination of the shared work tasks, knowledge, skills, and attributes required to perform a range of job functions across a group of related real-life occupations. In practice, it symbolizes a goal for education and training designed for workers who would be able to perform a broad variety of work tasks suitable to a large cluster of occupations.

To develop this learning occupation around computational thinking, she called for a process that would hold job-analysis workshops, validate the information, develop performance criteria and assessment guidelines, develop notional skills standards, and then validate them in an integrated skill standards model. Such a standards model would include content, assessment criteria, and measures of what people need to know and do to qualify for beginning-level employment. The model also contains an illustrative scenario of a routine work situation and a likely anticipated problem or breakdown.

Malyn-Smith contended that for computational thinking to get traction in the K-12 education community, it needs to be connected to frameworks and standards that are already implemented nationwide. An analysis of the Information Technology Career Cluster Initiative's model, for example, provides a way to organize a hierarchy of skills and knowledge that can be repurposed to support the integration of computational thinking in the K-12 arena. At the most basic level, this information technology skills framework calls for literacy and the ability to use common technology applications. Further up the hierarchy is fluency with information technology, which involves core knowledge and skill sets of technology-enabled workers employed in any industry sector. At the highest level of this model are the skill sets necessary for IT producer or developer careers—those that involve the design, development, support, and management of hardware, software, multimedia, systems integration, and services.

Malyn-Smith proposed that this hierarchy could be adapted for an appropriate learning progression in computational thinking. She suggested that grades K-4 might be devoted to computational thinking literacy, career awareness, and computational skills for learning. Grades 5-8 would also focus on computational thinking literacy but would fold in career exploration and learning about computational thinking skills for various STEM careers. Grades 9-10 would address computational thinking for all careers—students could explore and experiment with computational thinking in a variety of different contexts. Grades 11-12 would focus on providing pathways to college and careers, especially those for which competence in computational thinking (and computer science and engineering) will confer significant advantages. Postsecondary education and training would separate into two tracks—specialized computational thinking skills and competencies that are useful for STEM professions, and the use of basic computational thinking applications and tools for professions in all career tracks and in all industries.

Finally, Malyn-Smith noted that the Department of Education has developed a number of career clusters organized around a similar framework. Each cluster model includes a core skill set called "IT applications" to which computational thinking concepts and ideas can be attached.

Understanding this organizing framework helps teachers visualize the skills progression that leads learners from technology literacy to technology careers. Once computational thinking is better defined and examples are developed to illustrate what it looks like in practice, a similar model might be developed to help educators and other stakeholders visualize the K-adult skills progression of computational thinking from computational thinking literacy to STEM careers. Because our national community of educators already recognizes this framework, they will be more inclined to accept and integrate computational thinking into their programs and curricula. The tools and resources developed to facilitate other programs using similar models can then be adapted to support the integration of computational thinking nationwide.

4.4.6 Jan Cuny, National Science Foundation, CS 10K Project

Jan Cuny of the NSF's Broadening Participation in Computing Initiative discussed the CS 10K project, whose goals are to develop a new high school curriculum in computing and then to insert this revised curriculum so that it is taught in 10,000 schools by 10,000 well-prepared teachers by 2015.

The project focuses primarily on high schools because high schools have very little computer science education today. Cuny noted that without the high school piece, anything done at middle school will be lost and anything done at the college level will be insufficient. She further pointed out that of the few high schools that do offer "computer science" education, most do so by focusing on the vocational track and skills like keyboarding rather than deep computer science abstractions and so on. Last, she argued that the number of students who initially pursue computer science majors in college usually reflects the number of students who later graduate with a degree in the discipline. To increase the pool of computer scientists, it is necessary to provide high school students with opportunities for computer science education so that they enter college already interested in the discipline.

For the CS 10K project, the AP course for computer science is central. AP courses are in high demand in the nation's high schools, even if these schools often resist adding new courses. Further, the AP program is the primary point of national leverage—rather than going school district by school district to win approval for a computer science curriculum, one can simply invoke the AP computer science standard. Cuny also hopes that the new AP computer science courses will be an impetus for college curriculum reform, much as revisions to the calculus AP test helped drive changes in university teaching of calculus.

The CS 10K project seeks to develop courses that are engaging, accessible, inspiring, rigorous, and focused on the fundamentals of computing and computational thinking. As for content, a set of computational thinking practices are integrated with material built around the computing themes of big ideas, critical concepts, and enduring understanding.

Cuny also pointed to the need for feeder courses to AP programs in high school. She proposed that introductory courses in computing could be built on what schools already teach about computer science. Cuny described one example of such a course in the L.A. Unified School District (LAUSD)—the Exploring Computer Science (ECS) course, taught by Jane Margolis since 2008, and currently taught to about 900 students in 20 schools across Los Angeles.[33] In California, this course receives a general elective ("G") credit, which makes it eligible for college-prep credit.

Cuny reported that this course has generated significant interest from educators around the country and that there have been a number of requests for teacher training for this course. LAUSD has also created a mentoring and coaching program for computer science teachers because they are almost always completely isolated and benefit from having some outside reinforcement as well.

Cuny highlighted a number of university, non-profit, and industry partnerships, including a LAUSD-UCLA-Google partnership, a Georgia Tech-Wayne partnership exploring certification, and the NSF-UTEACH effort combining education majors with STEM majors. In business schools, Georgia Tech is also looking at certifying experienced information technology workers and pairing them up with a teacher in the business school. The University of Delaware-Chester School District project paired Chester District schools up with a service learning group at the University of Delaware that sends graduate students into classrooms to help teachers and kids use laptops in the classroom. Prior to this program, the district's laptop-for-every-child effort had resulted in hundreds of laptops stacked in school closets because the teachers did not have the training to use them. Finally, Cuny mentioned the National Lab Day project, which works to connect scientists, including computer scientists, with classroom teachers.

In addition to the curriculum development component, Cuny noted other challenges as well, such as getting the new curriculum into the schools, teacher preparation and ongoing professional development, and so on. She particularly called attention to the current patchwork of state standards, credit issues, and certification requirements—in her words, "they are a mess." Cuny and her colleagues are working with the Association for Computing Machinery (ACM) Education Policy Committee,

---

[33] Exploring Computer Science (ECS), website, http://www.exploringcs.org/. Last accessed February 7, 2011.

the ACM Education Policy Council, and the Computer Science Teachers Association to address some of these standards and certification issues.

Cuny summarized the challenges of introducing rigorous computer science education in high school as follows:

- We need computing classes at the local school level.
- We need standard certification and credit decisions at the state level.
- We need universities to step up and say that they will give credit for these courses.
- We need universities to step up and add computer science to their preferred list of courses for high school applicants.

Last, Cuny said she did not believe that computing and computer science do not fit well into current STEM education initiatives. She noted that as difficult as it is to train teachers who are already teaching computer science to teach even more computing to an increasingly rigorous standard, training science teachers who have little or no incentive to do so is even harder. In the long run, there is value in integrating computing into STEM education, but for now the CS 10K project serves as a kind of discipline-specific "race to the top."

4.5 EDUCATING THE EDUCATORS

4.5.1 Questions Addressed

- What are the goals for teachers and educators to bring computational thinking into classrooms effectively? What milestones do we hope to reach in computational thinking education?
- How should training efforts, support, and engagement be adapted to the varying experience levels of teachers such as pre-service, inducted, and in-service levels?
- What approaches for computational thinking education are most effective for educators teaching at the primary versus middle school versus secondary level? What methods might best serve the generalist teaching approach (multisubject/multidiscipline)? What method might best serve subject specialists?
- How does computational thinking education connect with other subjects? Should computational thinking be integrated into other subjects taught in the classroom?
- What tools are available to support teachers as they teach computational thinking? What needs to be developed?

Participants:
> Michelle Williams, Michigan State University
> Walter Allan, Foundation for Blood Research, EcoScienceWorks Project
> Jeri Erickson, Foundation for Blood Research, EcoScienceWorks Project
> Danny Edelson, National Geographic Society

Committee respondent: *Larry Snyder*

4.5.2 Michelle Williams, Michigan State University

Michelle Williams of Michigan State University discussed her experiences in facilitating teacher professional development in support of computational thinking-based science education.

Project and Curriculum

Williams and her colleagues originally set out to explore understanding by students in grades 5-7 of genetic inheritance of traits through an NSF career grant for a project titled "Tracing Children's Developing Understanding of Heredity over Time." The project curriculum was developed under the Web-based Inquiry Science Environment (WISE) instructional framework and aligned with the state and national science standards set forth in a number of works,[34] and it has been adopted by the school district in which the project has been operated.[35]

Williams argued that learning about genetic inheritance in middle school is a particularly interesting prospect because, although there is ample research at the secondary level indicating that students have many non-normative ideas about the topic, research is needed on middle and upper elementary school students' understanding of genetics concepts.[36]

---

[34] "STEM Education Statements and Letters," website, American Association for the Advancement of Science, available at http://www.aaas.org/spp/cstc/docs/09_06_02education.pdf, last accessed May 23, 2011; Minnesota Department of Education, "2007 Minnesota Mathematics Standards and Benchmarks for Grades K-12," website, Minnesota Department of Education, available at http://education.state.mn.us/MDE/Academic_Excellence/Academic_Standards/Mathematics/ index.html, last accessed May 23, 2011; NRC, 1996, *National Science Education Standards*, Washington, D.C.: National Academy Press. Available at http://www.nap.edu/catalog.php?record_id=4962. Last accessed February 7, 2011.

[35] Because of a recent change in the state curriculum standards that requires students to learn ecology in the sixth grade, Williams and her colleagues had to adjust their planned curriculum to teach ecology in addition to genetic inheritance.

[36] Elizabeth Engel Clough and Colin Wood-Robinson, 1985, "Children's Understanding of Inheritance," *Journal of Biological Education* 19(4):304-310; Colin Wood-Robinson, Jenny Lewis, and John Leach, 2000, "Young People's Understanding of the Nature of Genetic Information in the Cells of an Organism," *Journal of Biological Education* 35(1):29-36; Dennis

For example, many students have difficulty understanding the contributions of both parents to the genetic makeup of their offspring.[37] Students also have trouble understanding the concept of cells, for example, the structure and functions of cell organelles related to heredity.[38] Finally, students often conceptualize gene and trait as being equal, while being unable to distinguish between genotype and phenotype.[39]

In the fifth- through seventh-grade sequence, students are expected to learn to understand concepts related to cell structure, cell function, mitosis, and biological reproduction (both sexual and asexual), and the notion that heredity is the transmission of genetic information from one generation to the next. For example, fifth-grade students are tasked with investigating why organisms have similar and different features. Seventh-grade students carry out more sophisticated investigations, such as studying Mendel's law of segregation and using scientific evidence to make claims about the genotype and phenotype of an unidentified parent. At each level, the project provides scaffolding to help students learn how to use evidence to write better scientific explanations.

In Williams' project, students use animations and visualizations to understand abstract concepts. For example, they use simulations of mitosis to understand phases of cell division, Punnett squares to determine the genotypes and phenotypes of different generations of plants, and the Audrey's Garden animation to make distinctions between inherited and acquired traits.

The project curriculum calls for substantial collaboration between students and teachers. Some of this collaboration comes in the form of training videos of other teachers who have been involved in WISE in general in other places, showing how they work in their role or how they use the computer as a partner, and so on. Collaboration also occurs during small working group sessions during teacher training. In the classroom, stu-

---

Borboh Kargbo, Edward D. Hobbs, and Gaalen L. Erikson, 1980, "Children's Belief and Inherited Characteristics," *Journal of Biological Education* 14:137-146; and Grady Venville, Susan J. Gribble, and Jennifer Donovan, 2005, "An Exploration of Young Children's Understandings of Genetics Concepts from Ontological and Epistemological Perspectives," *Science Education* 89:614-633.

[37] Colin Wood-Robinson, 1994, "Young People's Ideas About Inheritance and Evolution," *Studies in Science Education* 24:29-47.

[38] Enrique Banet and Enrique Ayuso, 2000, "Teaching Genetics at Secondary School: A Strategy for Teaching About the Location of Inheritance Information," *Science Education* 84(3):313-351; Jenny Lewis and Colin Wood-Robinson, 2000, "Genes, Chromosomes, Cell Division and Inheritance—Do Students See Any Relationship?" *International Journal of Science Education* 22(2):177-195.

[39] Jenny Lewis and U. Kattmann, 2004, "Traits, Genes, Particles and Information: Revisiting Students' Understandings of Genetics," *International Journal of Science Education* 26:195-206.

dents work in pairs and the instructors consult with the students on their work, encouraging them to reflect on their learning by asking questions.

Teacher Professional Development

Williams and her team have invested considerable effort in teacher development. Research supports the proposition that teacher experience and content knowledge are important factors influencing student learning outcomes.[40] Furthermore, state and federal educational standards have increased teacher accountability through increased standardized testing and assessment.

To provide sustained professional development for in-service teachers, the project includes half-day sessions for professional development, for which participating teachers receive release time; summer workshop sessions; and after-school professional development meetings. Through these development sessions, teachers can collaborate across grade levels to think about curriculum coherence. They are also able to access science materials and learning technology such as the Wisconsin Fast Plants[41] and the Audrey's Garden programs. Using the WISE Genetic Inheritance Curriculum, teachers in professional development can also think about how to integrate various computational models (e.g., simulations) into their teaching of genetics. They learn how to analyze students' online work through embedded assessments and across-grade assessment items. Finally, teachers involved in this project interface with an instructional model in the curriculum that scaffolds students in using evidence to support claims.

Williams noted that some of her teachers were to some extent intimidated by the technology used to teach various concepts. She argued that

---

[40] See, for example, Hilda Borko, 2004, "Professional Development and Teacher Learning: Mapping the Terrain," *Educational Researcher* 33(8):3-15; Jodie Galosy, Jamie Mikeska, Jeffrey Rozelle, and Suzanne Wilson, 2008, "Characterizing New Science Teacher Support: A Prerequisite for Linking Professional Development to Teacher Knowledge and Practice," paper presented at the American Educational Research Association Annual Meeting, New York, March 2008; Suzanne Wilson and Jennifer Berne, 1999, "Teacher Learning and the Acquisition of Professional Knowledge: An Examination of Research on Contemporary Professional Development," *Review of Research in Education* 24(1):173-209; NRC, 1996, *National Science Education Standards*, Washington, D.C.: National Academy Press, available at http://www.nap.edu/catalog.php?record_id=4962, last accessed February 7, 2011; NRC, 2006, *Taking Science to School: Learning and Teaching Science in Grades K-8*, Washington, D.C.: National Academy Press, available at http://books.nap.edu/catalog.php?record_id=11625, last accessed February 7, 2011.

[41] "Wisconsin Fast Plants," website, University of Madison-Wisconsin, http://www.fastplants.org/. Last accessed May 23, 2011.

some level of discomfort was to be expected but that to mitigate this concern, she and her colleagues undertook several activities:

- They engaged in discussions with teachers from various grade levels across the district about the technology used.
- They spent significant amounts of time allowing teachers to use and interface with the technology as if they were students. Teachers reflected on and engaged with particular technologies, such as animations or some other types of visualizations, so that they would become accustomed to what their students would be doing. In this context, such reflection helps teachers to think about ways to make the subject clearer to students.
- They enlisted the assistance of teachers who were comfortable with technology to coach the technology-apprehensive teachers.

Williams also suggested that anxiety in one area can be conflated with anxiety in another. In one example, Williams explained that she worked with a former English teacher who had recently moved to science education. The teacher's discomfort with the technology associated with the project seemed to stem more from "the fact that she doesn't feel as confident with the science in general, and . . . technology kind of increases her anxiety."

Williams discussed the challenges of sustaining professional development activities, which often end when development grants and other outside funding end. This fate did not befall Williams' project, because the school district itself saw enough value in her program to adopt the project curriculum. She also pointed out that she built teacher support through open communication of outcomes with teachers. And she acknowledged the value of community support, for example, from parents who say they feel that their children are really learning and are excited about studying.

Williams drew several key conclusions from this project:

- It is critical to build relationships with key stakeholders that include principals, administrators, and parents, as well as the community at large.
- Teachers need adequate time to reflect on their teaching.
- Teachers-student collaborations can reduce anxiety for teachers and students.
- Teachers can enhance their own content knowledge and pedagogical knowledge by engaging in conversations about the curriculum and focusing on what their students do as well as what their students learn.

4.5.3 Walter Allan and Jeri Erickson, Foundation for Blood Research, EcoScienceWorks Project

The EcoScienceWorks project was supported by the National Science Foundation ITEST (Information Technology Experiences for Students and Teachers) program to develop a computer-based curriculum and inquiry-based tools to teach ecology and environmental science topics in curricular units that also introduce basic computer modeling. Under the terms of the request for proposals for ITEST projects, the goal was to integrate programming into an existing curriculum but not to add additional content.

The teachers who participated in the EcoScienceWorks development effort were familiar with the use of activity-based lessons, and they were quite enthusiastic about the opportunity to work with educators from Maine Audubon and collaborate and develop field exercises. But they were also interested in using their laptops to better support their student learning. As one of them said, they no longer felt they needed to be "the sage on the stage," and they were ready to have a more student-centered classroom. Teachers were tasked with writing a unit and lesson plans that would be built around the computer simulations and also with designing a field exercise that would go along with each of the simulations that were part of the software.

The development effort approached the use of programming subliminally, rather than as the primary focus of the effort. Downplaying the use of programming responded to the developers' concern that some teachers might rebel because the Maine learning standards did not include programming, and thus it would be difficult to justify spending scarce classroom time teaching programming.

Instead, the effort emphasized the development of ecology simulations. Five different simulations, all of them based on Maine habitat and featuring ecology content that the teachers were already teaching in their classroom, were developed. After field-testing these simulations (and the corresponding unit and lesson plans), teachers reported that this simulation-based approach enabled them to teach ecology in their middle school science classrooms more effectively. With this experience behind them, the teachers were more receptive to teaching programming.

Allan and Erickson illustrated their work by focusing on a simulation called Runaway Runoff. In this simulation, students conduct experiments on phosphorus pollution using a simulated lake ecosystem. By collecting and graphing data, they discover the connections between phosphorus level, algae growth rate, decomposition rate, and oxygen depletion, ultimately illuminating the ecological concept of eutrophication.[42]

The Runaway Runoff simulation depicts a lake ecosystem, with fish, zooplankton, and algae that are visible to students and bacteria that are invisible to students. In the first experiment, the simulation challenges the students to develop a food web by examining the contents of the digestive tracts of the trout and zooplankton.

In the second experiment, students examine decomposing algae in the lake ecosystem. Specifically, they change the level of phosphorus coming into the ecosystem and see how changes in phosphorus level affect the algae population in the lake and the concentration of dissolved oxygen. The students also learn that there are unseen organisms in this lake—the bacteria that act as decomposers.

In the final experiment, students are asked to predict the impact of increasing levels of phosphorus on the different populations of fish and zooplankton. Initially, students might predict that increasing levels of phosphorus result in increasing levels of algae, which in turn can support increased levels of zooplankton and thus, it would seem, an increase in the trout population. And so they might be surprised when they see that increasing the phosphorus levels leads to declining levels of trout. However, by running the simulations and observing the levels of dissolved oxygen, they can see that as algae increase, the bacteria use up some of the oxygen as they decompose dead algae, thus reducing the viability of the environment for the trout.

The Runaway Runoff simulation enables a cognitive cycle to occur. Students make a prediction; use the simulation for testing, tinkering, and playing; observe what happens; refine their mental model of how the system works; and then make further predictions. On the basis of essays written or posters created by students that describe how runoff affects lake ecology, Allan and Erickson believe that students learn to make fairly sophisticated mental models of the lake ecosystem.

Allan and Erickson also described the "Program a Bunny" environment. In this environment, the bunny is an agent that the student programs to find and eat carrots in a field. The environment is also probabilistic, so that carrots are not always located in the same places in the field, and thus a successfully programmed bunny must account for a degree of randomness in its environment. Students can test different programming strategies in a number of increasingly complex scenarios.

Program a Bunny is supplied with some initial programming and

---

[42] A sample student worksheet from the project can be seen at "Runaway Runoff Exercise 1: Who's Who," Worksheet, available at http://simbio.com/files/EBME_WSExamples/RunawayRunoff_WkSh1_example.pdf. Last accessed February 7, 2011.

will run "out of the box." But the initial programming is, by design, inadequate for bunny success. Thus, students must learn to modify the program. Modification of the program initiates a cognitive cycle similar to that of the Runaway Runoff simulation—the student observes what happens to the bunny's success in finding carrots, develops a mental model of how the program works, and then thinks of another modification that is intended to further improve the bunny's performance.

Allan and Erickson also reported that students found it helpful to use a concrete representation of the bunny field—a tarp laid on the floor and marked off with a grid. Students would go back and forth between the tarp and the Program a Bunny simulation, and thus develop a better understanding of the rules needed for programming the bunny.

In the views of Allan and Erickson, the common theme between these two examples—ecology and programming—is that students can see that there are computational rules and logic underlying both environments. They believe that learning ecology through the use of agent-based simulations combined with an agent-based programming challenge provided their middle school students with a rich learning environment for computational thinking.

4.5.4 Danny Edelson, National Geographic Society

Danny Edelson oversees the National Geographic Society's broad-based efforts to improve geographic education in the United States and around the world.[43] He characterized the efforts as building "geo-literacy," the ability to reason effectively about far-reaching decisions—the decisions that affect other people and places that members of 21st century society routinely face. Geo-literacy requires an understanding of how Earth's interconnected human, ecological, and geophysical systems function, and the ability to apply that understanding to decision making in personal, professional, and civic settings.

Edelson focused on geo-literacy in his presentation, arguing that the skills required for geo-literacy have substantial overlap with those needed in computational thinking. This overlap is rooted partly in the fact that both geography and computer science are disciplines that promote, indeed require, systems thinking.

Specifically, Edelson argued that geo-literacy is essentially a systems view of the world—an understanding of the world as a set of interconnected human and social systems and physical environmental systems (requiring an understanding of both of these elements as systems and

---

[43] "National Geographic Society Education," website, National Geographic Society, http://education.nationalgeographic.com. Last accessed February 7, 2011.

how they interact with each other) and then the ability to apply this understanding in context for the consequential tasks that citizens and workers need to be able to perform in the world.

Edelson stated that he found it is very hard "to disentangle this kind of systems view of the world and geographic reasoning from computational thinking." He went on to add, "I generated several different questions that I thought were all fascinating for which I have no answers at all. But I would like to use geo-literacy as a case that would be similar to lots of other science, natural science, social science, or other STEM disciplines."

Edelson posed a number of questions at the geo-literacy/computational thinking interface:

- What is the relationship between this concept of geo-literacy and computational thinking?
- To what extent is this systems view of the world a form of computational thinking in the way it actually plays out in the practice of geography or science or environmental science?
- How are these two things supportive of each other?
- How does computational thinking contribute to development of geo-literacy, and how does an understanding of Earth's systems contribute to development of computational thinking?
- How, if at all, might being an underdeveloped computational thinker impede learning of geo-literacy, and vice versa?
- How, if at all, might sophisticated computational thinking actually somehow interfere with developing an understanding in a scientific discipline like geography?

Edelson spoke of some of the issues that arise in understanding geographic data. For example, geographic data look continuous on Earth's surface viewed from afar, but in fact are pixilated when viewed close up. The actual physical situation on the surface is represented by continuous data, but the instruments of the geographer can represent the data only as pixels with rigid and discontinuous borders between them. Edelson reported a conceptual change in students when they go from viewing a map as a continuous representation to understanding a map as being a representation of discrete pixels or cells, with all the positive and negative implications of that for what they actually want to do with the data. (Pixilation means that it is impossible to determine from the data any reading that requires a smaller bin size, such as the ambient temperature on one's birthday.)

Another issue arises with maps that use different colors to represent different temperatures. Although it makes physical sense to subtract two temperatures (e.g., January's temperature from July's temperature), it does not make much sense to subtract yellow from red. That is, representations on the map cannot be manipulated in the same way as the underlying physical parameters. Edelson found that students could understand this paradox by considering the idea that maps are actually regular arrays of numerical data being represented pictorially. This conceptual step enabled students to understand what it might mean to "subtract January's temperature from July's temperature" and when they might want to perform that operation.

In performing analytic work, Edelson noted the computational overtones of working with sets, doing queries, and understanding Boolean logic and Boolean operations. For example, a student might be asked to find counties in the United States whose African American population exceeds the Caucasian population. Such operations are critical to being able to analyze much geographic data effectively.

Sometimes, such operations go beyond manipulating logical relationships but involve set or spatial combinations. For example, a student might want to say, "I've got two regions that are outlines on a map; find me the intersection of those two regions," or, "I have a list of cities that meet one criterion and a list of cities that meet another criterion; show me the intersection of those two lists." Managing these operations intellectually calls for thinking about them as combinations in one sense and as spatial entities in another sense. That is, this kind of geo-literacy requires students to understand spatial relationships as analogous to a set.

Edelson noted that many problems of interest to the GIS (geographic information system) community involve constraint satisfaction, sometimes with multiple constraints. For example, Edelson and colleagues developed a high school environmental science course in which one of the challenges was to find an appropriate location for a coal-burning power plant in a region of Wisconsin. Students understood that one requirement for a large power plant is the nearby availability of a sufficiently large body of water to provide cooling for it. So they can use the query capability to identify the large lakes in that region. A second consideration is adequate proximity to some mode of transportation that will allow coal to be transported to the power plant. So they need the ability to find the regions that are close to railroads (also known as buffers). And then they need to combine the two requirements—the power plant must be close to a large body of water and to a railroad. In general, solving problems that involve satisfying multiple constraints requires algorithmic thinking.

Edelson closed his presentation by arguing that Earth models are best understood in terms of dynamic and spatial models. He illustrated the point by discussing a NetLogo model for infiltration and runoff processes in a region in the presence of precipitation. Dynamic simulations demonstrate where the water runs off, and a student can determine the amount of water running off a given point in space at various periods of time. Modifying the runoff processes is necessary to demonstrate the effects of different land use conditions (e.g., a developed community has a surface that is much less permeable, and thus more water runs off, and students see dramatically higher and more rapid runoff showing up in that scenario).

In discussion, Mike Clancy suggested that the causal relationships depicted in these models are similar to the causal relationships entailed in understanding what a computer program actually does in execution, so that, for example, a student needs to understand what causes a program bug or a program to perform in a certain way.

Robert Panoff noted the importance of understanding limitations in the underlying data. In response, Edelson said that in his view, the issue of discrete versus continuous data is a placeholder for the whole issue of data quality, where it comes from, what you can and should be doing with it, and how you question it. He went on to say that anomalous data often catches people's attention and gives them an opportunity to see what's going on. He illustrated the point with an example taken from 1992-1993 when he was working with data provided by the National Center for Supercomputing Applications. This data set indicated a very strange warm spot in Europe, and all of the students noticed and asked about it. It turned out to be an anomaly in the models that generated the data or in the device. Many people argued for cleaning up that data so that the anomaly would not show up, but Edelson said that the anomaly was pedagogically valuable because it provided an important teachable moment.

Uri Wilensky asked about the importance of students collecting data themselves and then using that data to try to fit models to that data, rather than using data provided by others. Edelson concurred about the importance of collecting data, but said that he was not sure that data collection itself was computational in nature. He also said he was prepared to rethink that assertion.

4.6 MEASURING OUTCOMES (FOR EVALUATION) AND COLLECTING FEEDBACK (FOR ASSESSMENT)

4.6.1 Questions Addressed

- How can learning of computational thinking be assessed?
- What tools are needed to assess learning of computational thinking knowledge and capabilities? Which are available? What needs to be developed?

- What roles should embedded assessments play? What other assessments are needed?
- How can capabilities and skills of individuals be assessed when students are working collaboratively?
- How should the education community measure the success of its efforts? How can we compare the strengths and weaknesses of different efforts?
- What can be learned from efforts currently underway, and from efforts in our country and in other countries?

Participants:
Paulo Blikstein, Stanford University
Christina Schwarz, Michigan State University
Mike Clancy, University of California, Berkeley
Derek Briggs, University of Colorado, Boulder
Cathy Lachapelle, Museum of Science, Engineering is Elementary Project

Committee respondent: *Janet Kolodner*

4.6.2 Paulo Blikstein, Stanford University

Paulo Blikstein is an assistant professor of education and (by courtesy) computer science at Stanford University. His presentation discussed implementations of computational thinking and computer-based model-building activities within the context of a real undergraduate materials science/engineering classroom. He also shared some of his ideas for assessment of student learning under these circumstances.

Blikstein discussed "restructurations," a term that refers to the multiple ways of representing and encoding specific knowledge, each of which has different cognitive properties.[44] As a result, one representation might be more easily learned than another. The canonical example of a restructuration is that multiplying two numbers that are represented as Roman numerals is much more difficult than multiplying the same two numbers represented as Arabic numerals, even though each operation contains identical content.[45] The general lesson, Blikstein noted, is that "how we encode knowledge has a deep impact on how difficult it is to do things."

Blikstein demonstrated how to apply restructuration to understand-

---

[44] Paulo Blikstein and Uri Wilensky, 2010, "MaterialSim: A Constructionist Agent-based Modeling Approach to Engineering Education." In M.J. Jacobson and P. Reimann (eds.), *Designs for Learning Environments of the Future: International Perspectives from the Learning Sciences.* New York: Springer.

[45] This example was created by Uri Wilensky and Seymour Papert in recent work.

ing ideal gases in physics. In particular, he noted that the laws of ideal gases traditionally entail equations such as the Maxwell Boltzmann distribution and the relationship between pressure and volume. Blikstein offered an alternative restructuration based on computational thinking that represents a gas as a collection of molecules moving in a gas chamber governed by a simple rule: a molecule will move forward until or unless it bumps into another molecule or wall, at which point it will bounce back. This simple rule applied in this agent-based model results in aggregate behavior of the collection of gas molecules that is identical to that described by the formal gas law equations. Blikstein asserted that the computational representation of the gas laws is simpler and easier to learn than are the equations.

He also described how to reformulate a number of complex concepts from undergraduate-level materials science. Blikstein noted that students in traditional introductory materials science courses encounter new equations at a very rapid rate (one new equation every 150 seconds, not counting intermediate steps in a derivation). It is often that many different equations and models are needed to develop an understanding of a particular concept. These equations must be combined and manipulated to arrive at the final result.

Blikstein argued that an agent-based approach helps students to explore these complex and intertwined concepts more easily, and further that the rules and mechanisms governing the behavior of individual atoms can be used to understand a number of different crystal phenomena in materials science, such as growth, solidification, diffusion, and so on. An example of a relevant mechanism might be for molecules to "look around and see if they are surrounded by different neighbors or equal neighbors" and then cluster or disperse "based on their neighborhood." Similarly, solidification follows a comparable process except that an "atom in the liquid is kind of going around and looking for solid neighbors where it can attach itself."

Blikstein also described some of the challenges in assessing and giving objective feedback on open-ended projects with varying levels of complexity and explanatory power. These challenges included the following:

- How do we go about looking at various artifacts and understanding what students are doing?
- How do we assess the relative levels of complexity of the artifacts?
- How do we use assessment to provide feedback to students to improve their models as well as their understanding of concepts?

Blikstein described several tools to facilitate assessment—rubrics and maps, coding patterns over time, and representational shifts.

- *Rubrics and maps.* Based on the actual code that students generate, maps can be created that track the programming steps and decisions students were making. These maps capture many dimensions of students' decision making, such as how they define the system, how they define the rules of the system, how they define what the agents are doing, and so on. From these large maps, it is possible to categorize the rules embedded in the system and assess the sophistication of the rules. Evaluators can check each map to see if a student used various affordances of the programming language. For example, is this student using collisions? Is this student using neighborhood checking? Agents moving? Agents seeking agent clusters, walls, or energy? Blikstein argued that the greater the number of affordances appearing in a map, the more sophisticated the underlying model is likely to be, although this measure is not absolute and in many respects depends on the phenomenon being modeled.
- *Coding patterns over time.* Such patterns document how a student's code changes over time (e.g., what is added or deleted, what is found each time compilation is attempted). For example, one can count the number of characters in a program submitted for compilation. Some students—typically novice students—exhibit a pattern in which the code is more or less constant for several compilations but then jumps significantly in size. For other students (typically more expert students), there are fewer large increases in code size—code size increases more or less linearly over time. Blikstein asserted that such knowledge can be exploited to help tailor the most effective way to give feedback to different kinds of students.
- *Representational shifts.* Changes in how a student represents or depicts physical phenomena can indicate differences in the level of sophistication of his or her understanding. For example, Blikstein compared two groups of students, one that had been exposed to computational modeling and one that had not. Each group was asked to sketch the process involved in a scientific phenomenon different from the one they were modeling, such as the impact of a change in temperature. The students with computational modeling experience drew and described a mechanism showing the behavior of the atoms as the temperature changed. Students who were not exposed to the activity instead drew a graphical curve showing the aggregate behavior of the atoms as the temperature changed.

4.6.3 Christina Schwarz, Michigan State University

Christina Schwarz, an associate professor in the College of Education at Michigan State University, described her work with elementary and middle school students using scientific modeling and practices. The MoDeLS (Modeling Designs for Learning Science) project works to involve students in science through the use, revision, and creation of models. Although not explicitly focused on computational models, some of her work, Schwarz believes, may apply to the ongoing dialog about computational thinking.

In the context of her work, a model is an abstract, simplified representation of some phenomenon which could include but is not limited to computational representations. Models also include physical models and diagrammatic models. Modeling involves constructing a representation that embodies aspects of theory or evidence; evaluating that representation or testing it against empirical evidence and scientific theory; using it to illustrate, predict, and explain; and revising the representation.

Schwarz and her colleagues believe that the underlying concepts of modeling are powerful for sense-making and for communication in science. She further noted an overlap between modeling practice and computational thinking, particularly the ideas of abstracting and decomposing systems, testing the model against actual data, and so on.

Schwarz argued that models can make important aspects of science accessible by helping students to understand invisible processes, mechanisms, and components in phenomena. Models promote both subject-matter and epistemological understanding, and they develop systems thinking skills. Most importantly, models can generate predictions and explanations for scientific phenomena.

Schwarz walked through a generic MoDeLS curriculum sequence that would be given to students. The first step is for the researchers to provide some sort of anchoring phenomenon in a scientific context. For example, a fifth grade unit starts with a question like, Would you drink liquid that collected in a solar still?, and continues, "You can't test it, you don't want to drink it, because you might get sick, so you have to design an initial model that you can use to begin thinking through what is going on."

The unit then provides some discussion about the nature and purpose of models. Such dialog is essential to abstracting knowledge for transfer to other kinds of systems and contexts, and to motivate and support the kinds of skills and habits of mind essential to computational thinking.

The third element of the unit is an investigation of the subject phenomenon through data gathering and students' testing of their models. Students evaluate their models and discuss the criteria for evaluation. Evaluation is thus another strategy for teasing out modeling practice and scientific thinking.

Last, the unit introduces scientific ideas that students can use to revise their models again. Here, students often use visual, diagrammatic models and computer simulations in different ways. Students may look at simulations and then use some of those ideas from the simulations in their diagrammatic models. Model design, revision, and analysis occur in the context of a small scientific learning community (i.e., their classmates). The students debate data and concepts, as well as evaluate and peer-review each other's work. Finally, students develop a consensus model at the end of the unit and explore applying the model developed to other contexts that they care about.

Schwarz noted that different science disciplines use different aspects of modeling to explore scientific phenomena. For example, flow diagrams and process diagrams might be most appropriate for modeling relationships between components of a biological system. Most of the MoDeLS effort focuses on various aspects of physical science, but the group is looking at exploring modeling in other areas.

Schwarz uses a four-level learning progression to guide the interpretation of student activities. This progression is continually revised and improved based on their assessment outcomes.

- Level I focuses on students' reflections on their existing practices of modeling around the idea that children often begin modeling practice by drawing literal illustrations but have yet to really grasp the purpose of or use for models.
- Level II characterizes student use of models and shows that students are constructing and using models to illustrate and explain to an audience how phenomena occur. Although students at Level II are still somewhat literal, they are moving closer to the use of abstraction.
- Level III is even more sophisticated, as students move farther along the literal-to-abstract scale closer to the abstract end of the spectrum.
- Level IV students are constructing and using models spontaneously in a range of domains to help their thinking and problem solving. For example, students might be prompted to consider, before they test their model, how the world would behave. Schwarz argued that this fourth level is most similar to the types of modeling a computer scientist would do.

Schwarz also commented on assessment. Specifically, she noted that her assessments seek evidence in student work of engagement in modeling:

- Around different content knowledge for which students did not receive explicit instruction—to determine what aspects of modeling practices might be used across contexts.
- By applying their models to familiar and less familiar contexts. Schwarz described an example in which a student noted that she was actually applying the condensation and evaporation model to simple experiences at home like boiling water.

- Mapping between representations and the real world, as illustrated by students' application of their models in a specific context.
- Evaluating and revising their models for items like relevancy or saliency, evidentiary support, communicative power, and so on.

Schwarz and her colleagues use a variety of tools to obtain such evidence. Although there is some use of written pre-test and post-test items involving scientific modeling, they also use reflective interviews with students and in-person or videotaped observations of in-class student interactions. These qualitative instruments are designed to probe content that was both explicitly and not explicitly taught to examine transfer to other disciplines and the time evolution of student modeling practices and thinking.

Nevertheless, she was aware that their assessment efforts had a number of limitations. For example, many young students often see modeling and scientific thinking as a school-only activity that is unrelated to daily life rather than thinking of models as tools useful for their own purposes. Although they understand in principle the notion of evaluating each others' models according to relevant objective criteria, in practice they sometimes fail to do so in the classroom environment, instead deferring to the classmate they like better or the classmate who is the loudest.

Students also sometimes focus on the external audience when communicating through a model; that is, they may formulate their comments and responses based on what they think their teachers want to hear and what they think are "correct" answers, rather than what they themselves think.

Last, Schwarz noted that pedagogical constraints often result from the curricular and learning approaches determined by the various schools. As an example, Schwarz explained that in one school, the state-wide curriculum mandated that before fifth grade, science teachers are not to discuss phenomena at the cellular or atomic level because they are invisible. In response, Schwarz and her colleagues developed a special unit on evaporation and condensation that was actually an attempt to bridge the project's elementary learning goals to a particular state guideline prohibiting discussion of atoms.

4.6.4 Mike Clancy, University of California, Berkeley

Mike Clancy, from the Department of Computer Science at the University of California, Berkeley, addressed the topic of assessment for introductory programming classes. His top-level goals for students could be characterized as knowing when given aspects of computational thinking are applicable, when they are not applicable, and how these aspects are applied when they are applicable.

Clancy described two complementary approaches that are useful in assessment and evaluation. The first approach is based on case studies. A case study is an expert solution to a problem that is accompanied by a narrative of how that solution came to be. The expert, who may be a faculty member or a teaching assistant, provides a solution that addresses questions like why one approach to solving the problem was chosen over another and how problems originating in the first implementation of a solution were fixed (debugged).

Case studies are intended to make the expert's thinking visible to expose his or her design and development decisions. They demonstrate how abstract concepts are manifest in specific situations. They encourage reflection and self-monitoring, and they support collaborative learning and emphasize links among various problem solutions.

A typical problem might be to find the number of days spanned by two dates in the same year. (This problem arises in the third week of Berkeley's introductory programming course for non-majors, at which point they have been exposed to conditional programming structures such as "ifs" and how to deal with data but have not yet encountered recursion.)

One approach splits the solution into three situations—those in which the dates occur in the same month, those in which the dates occur in consecutive months, and those in which the dates are further apart. The first two situations are relatively easy to address, but the third is harder. Specifically, the solution for the third case depends on whether the months involved (including the intervening months, if any) have 28, 29, 30, or 31 days. Sometimes it is possible to kludge a solution when the dates are about 2 months apart, but if they are any further apart, a more systematic approach is needed.

At this point, the expert is faced with the question of crafting a solution to the third case that builds on the work already invested in crafting a solution to the first two cases. If one realizes that the day-span computation is essentially a subtraction of one date from another, a sensible approach is to change the representation of the dates involved into things that are easier to subtract—specifically, the date in month-day format is transformed into the number of days past January 1 for the year.

Using this idea, that is, finding a uniform representation for dates, students are then asked to address a number of related problems, such as computing the difference between two heights, finding the number of Saturdays spanned by two dates, and finding the number of days between dates in different centuries. In practice, their task is to understand the original solution (for the problem of computing the number of days between two dates in the same year) well enough so that they can modify the approach accordingly.

This case study also includes a debugging exercise; debugging is of course another key aspect of computational thinking. Imagine that the day-span program has been accidentally modified (e.g., one word is changed). Given the change in the output of the program as a starting point, students are asked to figure out what was changed and how to fix the problem.

The second approach used for assessment and evaluation involves lab-centric instruction, which emphasizes hands-on lab hours supervised by a teaching assistant rather than lecture and discussion. This instruction entails a variety of traditional programming tasks, such as writing, modifying, and analyzing a program. But because there is more lab time than in most lecture/discussion courses, the course also has room for a number of embedded assessment activities. For example, a lab period often starts with a quiz, and it provides opportunities for self-tests. "Gated collaborations" enable instructors to pose a question to students, and after any given student answers, s/he sees the answers of his or her lab mates.

In this environment, lab instructors can monitor most of what the students are doing and have a window into much of their thinking and not just their finished work. Thus, lab instructors can notice confusion when it occurs and address it immediately to provide targeted tutoring. The result is that instructors can nip confusion and misconceptions in the bud rather than having to wait for them to be revealed in some later venue.

4.6.5 Derek Briggs, University of Colorado, Boulder

Derek Briggs of the School of Education at the University of Colorado, Boulder, began by suggesting several questions that he believed should guide any assessment of computational thinking. His first question is, What is being assessed? A prerequisite for assessment is a common understanding of the important constructs and concepts of the topic being assessed. In the case of computational thinking, Briggs noted a lack of consensus on its essential elements and commented that even if one isn't willing to put down a thorough definition of what constitutes computational thinking, there has to be some common ground on the topic. What are the important elements?

Second, he argued for clarity about why the topic is being assessed. Briggs identified several possible reasons for assessing student understanding of a subject:

- *Evaluating a program.* If a pedagogical activity purports to promote student learning, the students involved in the activity must be assessed to see if the claimed learning indeed took place.

- *Grading of students.* In graded courses, a student's understanding of a topic often relates to the grade s/he receives.
- *Diagnosing a student's understanding of a subject in detail.* Pinpointing a student's misunderstanding of a particular subject-matter point provides feedback to a teacher about how to direct his or her pedagogical efforts to address that particular misunderstanding. For this particular application, multiple concepts of learning progression are helpful. A learning progression can be regarded as an ordered description of a specific student's understanding of a given concept as that student learns more about it; a description of successively more sophisticated understanding of a concept or ways of reasoning in a content domain; and also an ordered description of a typical student's understanding of a given concept as students learn more about it.
- *Developing a better intellectual understanding of a subject.* It sometimes happens that an attempt to assess a student's understanding of a subject demonstrates that the expert's understanding of the subject is incomplete, and it is through the act of developing an instrument, and developing questions for students that are intended to elicit information about the subject, that the expert gains insight as to what it is that the expert really meant.

Third, an instrument for the assessment must be appropriate to the purpose of the assessment. For example, if the purpose of the assessment is to grade students, an instrument may need only to record the percentage of correct answers provided by a student. However, if the purpose of the assessment is to diagnose a student's misunderstandings, the instrument must be constructed in a way that sheds light on the specific nature of those misunderstandings. Briggs also noted that diagnosing student misunderstandings does not necessarily entail open-ended interactions with students—carefully designed multiple-choice items can provide diagnostic information that is as meaningful as or more meaningful than that obtained through open-ended interviews.

Finally, Briggs argued for the importance of validating an instrument, contrasting the notion of validity to the notion of reliability. A valid instrument is one that accurately reflects a student's knowledge of the specific concepts of interest (i.e., what the investigator really wishes to assess), whereas reliability is concerned with the consistency with which an instrument can produce a given measure. He further noted that low reliability of an instrument was not necessarily problematic in the context of formative evaluations for real-time informing of in-class pedagogy or group-level comparisons.

4.6.6 Cathy Lachapelle, Museum of Science, Engineering is Elementary Project

Cathy Lachapelle, director of research and evaluation for the Engineering is Elementary (EiE) project at the Museum of Science, discussed her assessment and evaluation experiences with that project. EiE is a curriculum development and improvement effort that develops engineering guides and activities for children in grades 1-5.

Assessments of EiE activities are focused on what students learn and measure specific learning objectives.[46] Lachapelle noted that there is no existing standard "yardstick" against which to assess student learning about engineering. Thus, assessment efforts compare progress toward learning objectives in an EiE activity group to progress among students in a control group.

Lachapelle suggested that a variety of methods are available for assessing student learning, depending on the purpose of the assessment:

- *Class observation that focuses on collecting qualitative data.* Such data include information obtained from helping the teacher implement EiE, interviewing students to try to understand their attitudes with respect to the learning objectives, and observing how they perform against the learning objectives. To illustrate, Lachapelle noted that one of the learning objectives is to be able to reason from a model and understand that a model is representing something in the real world. During class observation, assessors talk to the teacher and students to see if the students are grasping the concept. (They might also point out different ways to structure the lessons so that students better understand the learning objectives.) A degree of uniformity in data collection is obtained by using the same standards and criteria in each observation.
- *Embedded assessments, which are often used by teachers to understand the pedagogical impact they are having on students as they go along.* Embedded assessment can be as simple as examining individual student performance on a particular worksheet, so that a teacher can better understand which students need more help, whether he or she should give clearer instructions, and so on.
- *Paper-and-pencil assessments, which are very difficult to construct but provide an excellent source of feedback.* EiE typically uses these paper-and-pencil assessments for summative evaluation. A great deal of work is involved in constructing assessments and testing them, piloting them, checking them for reliability, and then using them with hundreds of

---

[46] Not all investigation of student learning requires such objectives—specifically, some research is useful for understanding what students know in general and what they can do on average.

students. For example, developing multiple-choice questions that yield insight into student thinking is sometimes problematic. Lachapelle and her colleagues often ask students how they would answer a question, and unusual or incorrect student answers become alternative choices for answering the question. For example, Lachapelle said, "We asked kids what is the function of leaves in a plant and the kids said, to make food. We would say well why did you choose that answer? And they said because they make salad. You have learned that things are not always as they might seem or as you might expect." Ultimately, they discarded that particular question.

- *Performance assessments, which can be used either by teachers for their own understanding of what their students are learning (in formative evaluation) or by the curriculum development team as a summative evaluation of what students learned.* This type of testing is also time- and resource-intensive because the assessment must be administered and scored. EiE uses this type of assessment in the final project design exercise.

Speaking more broadly, Lachapelle addressed formative and summative evaluations in the EiE project. All work products require regular evaluation, including teacher guides, student exercises and activities, the learning goals, and teacher professional development materials and activities.

As is usually the case, formative evaluation is used to inform the development and improvement of products and processes. In the EiE context, formative evaluations seek evidence of growth in students' understanding and skills as stated in EiE learning objectives, determine the age-appropriateness of lessons and activities, and examine the ease of use of lessons and materials. Formative evaluation for EiE usually relies on feedback from teachers and students. Therefore, it is critical that researchers make sure that the lines of communication are open and that feedback received is considered in light of the project's set evaluation criteria. Lachapelle explained that if a researcher receives feedback that the project was great but too troublesome to clean up afterward or the standard for an activity was that a teacher be able to manage the activity the following year without any support staff, the activity would be revised accordingly.

The purpose of summative evaluation is to provide evidence to EiE stakeholders, including funders, school districts, teachers, and parents, that implementation of specific EiE activity is worthwhile. Robert Panoff was particularly struck by this concept of "being worthwhile" and argued that this concept is a key factor in terms of scaling, adoptability, and motivation for using the materials or the exercises. Lachapelle stated that one criterion for this type of evaluation is to show improved learning of target concepts among students as compared to a control group of students. The control group consists of students in a comparable classroom taught the same science and engineering topics but not with EiE curriculum materials and tools. In the ideal scenario, EiE has a large pool of teachers from which part are admitted to the EiE project and the other part remain as control groups. This process does not always work because of constraints of funding and time. Another example criterion is that teachers express increased efficacy and interest in teaching engineering to their students.

Randomized, controlled studies with external evaluators are the preferred method for evaluating and comparing efforts in education, said Lachapelle. NSF, for example, prefers this approach when seeking summative assessments in projects it funds. Unfortunately this type of assessment is very expensive to execute because there is usually a need for a fairly large number of students in order to randomize whole classrooms into different testing groups. Also external evaluators are an added cost and bear their own pros and cons. Although external evaluators are likely to be more objective in their assessments, they do not have the advantage of an ongoing relationship with the teachers, administrators, and students whom they are engaging and thus may miss subtleties that more familiar evaluators might observe.

In her discussion, Lachapelle cautioned that assessments and evaluations of computational thinking activities and materials require clearly specified learning objectives, which in turn require some community consensus regarding the content of computational thinking—that is, what is it that the community wants children at various ages to know (from early elementary school to college)? In the EiE context, some learning objectives include being able to identify a process, to explain what a process is in an engineering context, and to explain why the order of steps in a process is important.

She also argued that the learning objectives should align with psychological and developmental learning progressions, since doing so provides some guidance over time as to where students should be at each stage. Thus, learning objectives are and should be the object of research and design. She noted that EiE does extensive literature searches and local interviews with kids before beginning the design of each of its units in order to learn more about what kids know. For example, for a unit on sinking and floating, developers would do a literature search and then interview local students by asking them things like, "Do you know what it means to float?," "Do you understand why things float?," and so on.

Finally, Lachapelle commented that their assessments are also designed to address student attitudes toward science and engineering. Broadly speaking, these assessments indicate that girls tend be interested in engineering things when framed as helping to improve people's lives and boys tend to be interested in engineering things when framed in terms of constructing engineering artifacts.

5

Conclusion

As noted in the Preface and in Chapter 1, this set of two workshops was not intended to develop or to advance a consensus view of computational thinking. In both workshops, participants expressed a wide variety of views regarding both the nature of and the pedagogy for computational thinking. It is the committee's hope that the summaries from both workshops will help to stimulate in the relevant communities the creative thinking that is necessary for a consensus view on this topic to emerge in the future.

Appendixes

A

Workshop Agenda

PEDAGOGICAL DIMENSIONS OF COMPUTATIONAL THINKING
KECK CENTER, NATIONAL ACADEMIES, WASHINGTON, D.C.

February 4, 2010

8:30 AM-8:45 AM	**Welcome** *Marcia Linn, University of California, Berkeley, Committee Chair* *Jeannette M. Wing, National Science Foundation*
8:45 AM-10:15 AM	**Panel 1—Computational Thinking and Scientific Visualization**

- What are the relevant lessons learned and best practices for improving computational thinking in K-12 education?
- What are examples of computational thinking and how, if at all, does computational thinking vary by discipline at the K-12 level?
- What exposures and experiences contribute to developing computational thinking in the disciplines?
- How do computers and programming fit into computational thinking?

- What are plausible paths and activities for teaching the most important computational thinking concepts?

Presenters:
Robert Tinker, The Concord Consortium
Mitch Resnick, Massachusetts Institute of Technology
John Jungck, Beloit College, BioQUEST
Idit Caperton, World Wide Workshop

Committee respondent: *Uri Wilensky*

10:15 AM-10:45 AM **Break**

10:45 AM-12:00 PM **Panel 2—Computational Thinking and Technology**

- What are the relevant lessons learned and best practices for improving computational thinking in K-12 education?
- What are examples of computational thinking and how, if at all, does computational thinking vary by discipline at the K-12 level?
- What exposures and experiences contribute to developing computational thinking in the disciplines?
- How do computers and programming fit into computational thinking?
- What are plausible paths and activities for teaching the most important computational thinking concepts?

Presenters:
Robert Panoff, Shodor Education Foundation
Stephen Uzzo, New York Hall of Science
Jill Denner, Education, Training, Research Associates

Committee respondent: *Yasmin Kafai*

APPENDIX A

12:00 PM-1:15 PM	**Working Lunch**—*Lou Gross, University of Tennessee (via teleconference)*
1:15 PM-2:45 PM	**Panel 3—Computational Thinking in Engineering and Computer Science**

- What are the relevant lessons learned and best practices for improving computational thinking in K-12 education?
- What are examples of computational thinking and how, if at all, does computational thinking vary by discipline at the K-12 level?
- What exposures and experiences contribute to developing computational thinking in the disciplines?
- How do computers and programming fit into computational thinking?
- What are plausible paths and activities for teaching the most important computational thinking concepts?

Presenters:
Christine Cunningham, Museum of Science, Engineering is Elementary Project
Taylor Martin, University of Texas at Austin
Ursula Wolz, College of New Jersey
Peter Henderson, Butler University

Committee respondent: *Marcia Linn*

2:45 PM-3:00 PM	**Break**
3:00 PM-4:30 PM	**Panel 4—Teaching and Learning Computational Thinking**

- What is the role of computational thinking in formal and informal educational contexts of K-12 education?
- What are some innovative environments for teaching computational thinking?
- Is there a progression of computational thinking concepts in K-12 education?

What are criteria by which to order such a progression? What is the appropriate progression?
- What are plausible paths to teaching the most important computational thinking concepts?
- How do cognitive learning theory and education theory guide the design of instruction intended to foster computational thinking?

Presenters:
Deanna Kuhn, Columbia University
Matthew Stone, Rutgers University
Jim Slotta, University of Toronto
Joyce Malyn-Smith, Education Development Center, Inc.

Committee respondent: *Al Aho*

4:30 PM-4:45 PM	**Break**
4:45 PM-5:00 PM	**Open Discussion** Moderator: *Herb Lin, CSTB Staff*
5:00 PM-5:25 PM	**Special Session—Update from Jan Cuny** *Jan Cuny, National Science Foundation*
5:25 PM-5:30 PM	**Wrap-up**
5:30	**Adjourn Day One Public Sessions**

February 5, 2010

8:30 AM-8:45AM	**Welcome and Housekeeping** *Marcia Linn, University of California, Berkeley, Committee Chair*
8:45 AM-10:00 AM	**Panel 5—Report-back on Homework Assignments**

Committee respondent: *Brian Blake*

10:00 AM-10:15 AM	**Break**
10:15 AM-11:45 AM	**Panel 6—Educating the Educators**

- What are our goals for teachers and educators to bring computational thinking into classrooms effectively? What milestones do we hope to reach in computational thinking education?
- How should training efforts, support, and engagement be adapted to the varying experience levels of teachers such as pre-service, inducted, and in-service levels?
- What approaches for computational thinking education are most effective for educators teaching at the primary versus middle school versus secondary level? What methods might best serve the generalist teaching approach (multisubject/multidiscipline)? What methods might best serve subject specialists?
- How does computational thinking education connect with other subjects? Should computational thinking be integrated into other subjects taught in the classroom?
- What tools are available to support teachers as they teach computational thinking? What needs to be developed?

Participants:
Michelle Williams, Michigan State University
Walter Allan, Foundation for Blood Research, EcoScienceWorks Project
Jeri Erickson, Foundation for Blood Research, EcoScienceWorks Project
Danny Edelson, National Geographic Society

Committee respondent: *Larry Snyder*

11:45 AM-12:45 PM	**Working Lunch**
12:45 PM-2:15 PM	**Panel 7—Measuring Outcomes (for**

Evaluation) and Collecting Feedback (for Assessment)

- How can learning of computational thinking be assessed?
- What tools are needed to assess learning of computational thinking knowledge and capabilities? Which are available? What needs to be developed?
- What roles should embedded assessments play? What other assessments are needed?
- How can capabilities and skills of individuals be assessed when students are working collaboratively?
- How should the education community measure the success of its efforts? How can we compare the strengths and weaknesses of different efforts?
- What can be learned from efforts currently underway, and from efforts in our country and in other countries?

Participants:
Paulo Blikstein, Stanford University
Christina Schwarz, Michigan State University
Mike Clancy, University of California Berkeley
Derek Briggs, University of Colorado, Boulder
Cathy Lachapelle, Museum of Science, Engineering is Elementary Project

Committee respondent: *Janet Kolodner*

2:30 PM-4:00 PM	**Discussion and Wrap-up**

- Committee members summarize their individual reactions
- Floor thrown open to other workshop participants for discussion

4:00 PM	**Adjourn**

B

Short Biographies of Committee Members, Workshop Participants, and Staff

B.1 COMMITTEE

Marcia C. Linn (*Chair*) is a professor specializing in education in mathematics, science, and technology in the Graduate School of Education at the University of California, Berkeley. She directs the National Science Foundation-funded Technology-Enhanced Learning in Science (TELS) center. She is a member of the National Academy of Education and a fellow of the American Association for the Advancement of Science, the American Psychological Association, and the Center for Advanced Study in Behavioral Sciences. Her board service includes the American Association for the Advancement of Science board, the Graduate Record Examination Board of the Educational Testing Service, the McDonnell Foundation Cognitive Studies in Education Practice Board, and the Education and Human Resources Directorate at the National Science Foundation. Linn earned a B.A. in psychology and statistics, and a Ph.D. in educational psychology from Stanford University.

Alfred V. Aho (NAE) is the Lawrence Gussman Professor of Computer Science and vice chair of undergraduate education for the Computer Science Department at Columbia University. Previously, he conducted research at Bell Laboratories from 1963 to 1991, and again from 1997 to 2002 as vice president of the Computing Sciences Research Center. Aho's current research interests include quantum computing, programming languages, compilers, and algorithms. He is part of the Language and Compilers research group at Columbia. He is widely known for his development of the AWK programming language with Peter J. Weinberger and Brian Kernighan (the "A" stands for "Aho") and for his co-authorship of *Compilers: Principles, Techniques, and Tools* (the "Dragon book") with Ravi Sethi and Jeffrey Ullman. He wrote the initial versions of the Unix tools egrep and fgrep. He is also a co-author (along with Jeffrey Ullman and John Hopcroft) of a number of widely used textbooks on several areas of computer science, including algorithms and data structures, and the foundations of computer science. He is a past president of ACM's Special Interest Group on Algorithms and Computability Theory. Aho has chaired the Advisory Committee for the Computer and Information Sciences Directorate of the National Science Foundation. He has received many prestigious honors, including the IEEE's John von Neumann Medal and membership in the American Academy of Arts and Sciences. Aho was elected to the National Academy of Engineering in 1999 for contributions to the fields of algorithms and programming tools. He earned his B.A.Sc. in engineering physics from the University of Toronto and his Ph.D. in electrical engineering and computer science from Princeton University.

M. Brian Blake is a professor of computer science and associate dean of engineering at the University of Notre Dame. His research interests include the investigation of automated approaches to sharing information and software capabilities across organization boundaries, sometimes referred to as enterprise integration. His investigations cover the spectrum of software engineering: design, specification, proof of correctness, implementation/experimentation, performance evaluation, and application. Blake's long-term vision is the creation of adaptable software entities or software agents that can be deployed on the Internet and, using existing resources, manage the creation of new processes, sometimes referred to as interorganizational workflow. He has several ongoing projects that make incremental progress toward this long-term vision. In addition, he conducts experimentation in the areas of software engineering education and software process and improvement to determine the most effective methods for training students and professionals to develop module systems that by nature are distributed. Blake has consulted for such companies as General Electric, Lockheed Martin, General Dynamics, and the MITRE Corporation. He has published more than 95 refereed journal papers and conference proceedings in the areas of service-oriented computing, agents and workflow, enterprise integration, component-based software engineering, distributed data management, and software engineering education. Blake's work has been funded by the Federal Aviation Administration, the MITRE Corporation, the National Science Foundation, DARPA, the Air Force Research Laboratory, SAIC, and the National Institutes of Health. He earned his bachelor's in electrical engineering and doctorate in information technology and computer science from George Mason University.

Robert Constable is the dean of the Faculty of Computing and Information Science at Cornell University. Formerly he was the chair of the Computer Science Department for 6 years. He also heads a research group in automated reasoning and formal methods in the Computer Science Department, where he is a professor. Constable is a graduate of Princeton University, where he earned his A.B. in mathematics and worked with Alonzo Church, one of the pioneers of computer science. He did his M.A. and Ph.D. work at the University of Wisconsin with Stephen Cole Kleene, a Ph.D. student of Church and another pioneer of computer science. Constable joined the Cornell University faculty in 1968. He has supervised more than 43 Ph.D. students in computer science. He is known for work in connecting programs and mathematical proofs that has led to new ways of automating the production of reliable software. This work is known by the slogan "proofs as programs," and it is embodied in the Nuprl ("new pearl") theorem prover. He has written three books on this topic as well as numerous research articles. Since 1980 he has headed a project that uses Nuprl to design and verify software systems, instances of which are still operational in industry and science. Currently he is working on extending this programming method to concurrent processes, realizing the notion of "proofs as processes." In 1999 he became the first dean of the Faculty of Computing and Information Science, a unit that includes the Computer Science Department, the Department of Statistical Science, the Information Science Program, and the Program in Computer Graphics. It also sponsors the undergraduate major and graduate specialty in computational biology.

Yasmin B. Kafai is a professor at the Graduate School of Education, University of Pennsylvania. In addition, she spent more than a decade on the faculty at the UCLA Graduate School of Education and Information Studies. As a learning scientist, she has researched and developed media-rich software tools and environments, most recently Scratch, together with researchers at the MIT Media Lab, that support youth in schools and community centers in becoming designers of games, simulations, and virtual worlds. As part of her policy initiatives, she wrote *Under the Microscope: A Decade of Gender Equity Interventions in the Sciences* (2004) and participated in the national commission that produced the report *Tech-Savvy Girls: Educating Girls in the Computer Age* (2000) for the American Association of University Women. She also briefed the committee that prepared the National Research Council report *Being Fluent with Information Technology* (National Academy Press, Washington, D.C., 1999). She received her hauptdiplom in psychology from the Technical University of Berlin in Germany and her D.E.U.G. in psychology from the Université de Haute Bretagne II in France. While conducting research at the Massachusetts Institute of Technology Media Laboratory, she received her Ed.D. in human development and psychology from Harvard University.

Janet L. Kolodner is a Regents' Professor in the School of Interactive Computing at Georgia Institute of Technology. Her research over the past 30 years has addressed a wide variety of issues in learning, memory, and problem solving, both in computers and in people. During the 1980s, she pioneered the computer method called case-based reasoning, which allows a computer to reason and learn from its experiences. The first case-based design aids (CBDA), such as Archie 2 for architecture, came from her lab. During the early 1990s, she used the cognitive model implied by case-based reasoning to address issues in creative design with the development of programs like JULIA (planned meals), Creative JULIA (planned meals with leftovers), IMPROVISOR (simple mechanical design), and ALEC (simulated Alexander Graham Bell's invention of the telephone). Later in the 1990s, she used the cognitive model in case-based reasoning to guide the design of a science curriculum for middle school. Learning by Design™ is a design-based learning approach and an inquiry-oriented project-based approach to science learning that has children learn science from their design experiences. Learning by Design curriculum units and embedded sequencing structures were integrated into a full 3-year middle-school science curriculum called Project-Based Inquiry Science Digging-In (It's About Time) and published in 2009.

Most recently, Kolodner's research uses what she learned in designing Learning by Design to create informal learning environments to help middle schoolers come to think of themselves as competent scientific reasoners through Kitchen Science Investigators (science in cooking), and Hovering Around (motion, airflow, and design of hovercrafts). Kolodner is the founding editor in chief of the *Journal of the Learning Sciences* and is a founder and first executive officer of the International Society for the Learning Sciences. She has headed up the Cognitive Science Program at Georgia Tech and headed an organization called EduTech in the mid-1990s whose mission was to use what we know about cognition to design educational software and integrate it appropriately into educational environments. She has a B.S. from Brandeis University in math and computer science and an M.S. and a Ph.D. in computer science from Yale University.

Lawrence Snyder is a professor of computer science and engineering at the University of Washington in Seattle (UW). Snyder's research has focused on parallel computation, including architecture, algorithms, and languages. He has served on the faculties of Yale and Purdue Universities and has had visiting appointments at UW, Harvard, MIT, Sydney University, the Swiss Technological University, the University of Auckland, and Kyoto University. In 1980 he invented programmable interconnect, a method to dynamically configure on-chip components, and a technology used today for field-programmable gate arrays. In 1990 he was co-designer of Chaos Router, a randomizing adaptive packet router. He was the principal investigator of the ZPL language design project, the first high-level parallel language to achieve "performance portability" across all parallel computer platforms. Snyder is the author of *Fluency with Information Technology: Skills, Concepts and Capabilities*, a textbook for non-techie college freshmen that teaches fundamental computing concepts; the book is in its third edition. With former Ph.D. student Calvin Lin (University of Texas, Austin), he has written *Principles of Parallel Programming*, published in 2008. Snyder was a three-term member of the Computer Research Association Board of Directors, developing a series of best-practice white papers. He chaired the NSF CISE Advisory Board as well as several CISE directorate oversight panels and numerous review panels. The two National Research Council studies that he has chaired produced influential reports—*Academic Careers for Experimental Computer Scientists and Engineers* (1994) and *Being Fluent with Information Technology* (1999). He served three terms on the NRC's Army Research Laboratory Technical Advisory Board. He serves on ACM's Education Board, has been general chair or program committee chair of several ACM and IEEE conferences, and he is a fellow of both the ACM and the IEEE. He received a B.A. from the University of Iowa in mathematics and economics and his Ph.D. from Carnegie Mellon University in computer science as a student of A. Nico Habermann.

Uri Wilensky is a professor of learning sciences and computer science at Northwestern University and holds appointments in the cognitive science program and in complex systems. He is the founder and current director of the Center for Connected Learning and Computer-Based Modeling and also a founder and member of the governing board of the Northwestern Institute on Complex Systems (NICO). His most recent projects focus on developing tools that enable users (both researchers and learners) to simulate, explore, and make sense of complex systems. His NetLogo agent-based modeling software is in widespread use worldwide. Prior to coming to Northwestern, he taught at Tufts University and MIT and was a research scientist at Thinking Machines Corporation. Wilensky is a founder and an executive editor of the *International Journal of Computers for Mathematical Learning*. His research interests include computer-based modeling and agent-based modeling, STEM education, mathematics in the context of computation, and complex systems. He is a recipient of the National Science Foundation's Career Award as well as the Spencer Foundation's Post-Doctoral Award. He has directed numerous NSF research projects focused on developing computer-based modeling tools and studying their use. Among these tools are multiagent modeling languages, such as StarLogoT and NetLogo; model-based curricula such as GasLab, ProbLab, NIELS, and BEAGLE Evolution; and participatory simulation toolkits such as Calc-HubNet and Computer-HubNet. The tools enable learners to explore and create simulations of complex phenomena across many domains of natural and social science and, through creating and exploring such simulations, to deepen their understanding of core scientific concepts. Many of these tools are also in use by researchers across a wide variety of domains, including the natural sciences, social sciences, business, and medicine. By providing a low-threshold language for exploring and constructing models, Wilensky hopes to promote modeling literacy—the sharing and critiquing of models in the scientific community, in education, and in the public at large. Wilensky did his undergraduate and graduate studies in mathematics, philosophy, and computer science at Brandeis and Harvard Universities and received his Ph.D. in media arts and sciences from the Massachusetts Institute of Technology.

B.2 WORKSHOP PARTICIPANTS

Walter Allan is medical director and consulting scientist to ScienceWorks for ME at the Foundation for Blood Research (FBR), a non-profit independent research and education institution in Scarborough, Maine. He was the first pediatric neurologist in Maine and was the director of both the Pediatric and the Adult Neurology Divisions at the Maine Medical Center, where his responsibilities included teaching medical and pediatric residents prior to coming to the FBR. His interests include the consequences of central nervous system injury in children and science education. He is the principal investigator on a National Institutes of Health Science Education Partnership Award that has developed a curriculum (BiomedicineWorks) that introduces evidence-based medicine to advanced high school biology classes and a National Science Foundation ITEST (Information Technology Experiences for Students and Teachers) grant that has developed a computer simulation-focused curriculum (EcoScienceWorks) to teach ecology and introductory computer programming in Maine's seventh- and eighth-grade laptop-equipped classrooms.

Paulo Blikstein is an assistant professor at Stanford's School of Education, he has a courtesy appointment in the Computer Science Department.

His research focuses on computational literacy, low-cost educational technologies for students in low-income settings, and STEM education. His work cuts across age groups, and he has worked extensively with inner-city students in developing countries such as Brazil, Mexico, Senegal, and Costa Rica, but also with undergraduates in elite U.S. institutions. His research tries to bring the most cutting-edge computational tools to the classroom, creating environments for students to engage authentically in advanced, deep scientific inquiry. He completed his Ph.D. at the Center for Connected Learning and Computer-Based Modeling at Northwestern University, earned a B.S. in metallurgical engineering and an M.Sc. in digital systems engineering from the University of São Paulo, Brazil (1998, 2001), and obtained an M.Sc. from the MIT Media Lab (2002), where he was also a visiting scholar (2003).

Derek Briggs is chair of the Research and Evaluation Methodology Program at the University of Colorado, Boulder, where he also serves as an associate professor of quantitative methods and policy analysis. His research agenda focuses on building sound methodological approaches for the valid measurement and evaluation of growth in student achievement. Examples of his research interests in the area of educational measurement include (1) characterizing the gap between validity theory and practice in the context of high-stakes standardized testing and (2) developing and applying psychometric models to assess learning progressions. He holds a B.A. in economics from Carleton College and a Ph.D. in education from the University of California, Berkeley.

Idit Harel Caperton is a pioneer in using new-media technology for cultivating creative learning, innovation, and globalization through constructionist learning theory. She founded the World Wide Workshop in 2004 to leverage her unique blend of award-winning research, social entrepreneurship, and leadership in new-media learning projects around the world. In 2006 the foundation launched the Globaloria.org network to implement ways of using social media technology and Web 2.0 tools to teach innovative game making to and build computational creative capacities in youth worldwide. Throughout the 1980s and 1990s, Caperton conducted breakthrough research at the MIT Media Lab that led to publishing the book *Constructionism* with Seymour Papert. Her book *Children Designers* received the 1991 Outstanding Book Award from the American Educational Research Association. In 1995, she founded MaMaMedia and launched MaMaMedia.com, ConnectedFamily.com, and Papert.org. Pioneering kids' Internet media activities, MaMaMedia established global distribution and advertising partnerships and won numerous honors, including the Computerworld-Smithsonian Award (1999), the Internet industry's coveted Global Information Infrastructure Award (1999), and the 21st-Century Achievement Award from the Computerworld Honors Program (2002). In 2002, she was honored by the Network of Educators in Science and Technology and MIT "for devotion, innovation, and imagination in science and technology on behalf of children and youth around the world." In 2007-2008 MaMaMedia activities were reprogrammed for One Laptop per Child (OLPC). Caperton has served on numerous advisory boards (for Harvard University, MIT, CU-ATLAS, CUNY, PBSKids, TIG, MEET, and Saybot), and she has been an adviser to commercial, governmental, higher education, and not-for-profit organizations on inventing, developing, and harnessing technology and innovative programs to transform education. Caperton holds degrees from Tel Aviv University (B.A., 1982), Harvard University (Ed.M., 1984; CAS, 1985), and MIT (Ph.D., 1988).

Mike Clancy has been on the University of California, Berkeley, computer science faculty since 1977. He is an active member both of the U.S. computer science education community and of the community of researchers who study the psychology of programming. In 2009 he received the ACM SIGCSE award for lifetime contributions to computer science education. His work, in collaboration with Marcia Linn in Berkeley's School of Education, spans a spectrum from exploration of student misconceptions through development of curriculum components and programming environment features to support integration of programming knowledge. Clancy and Linn have focused in particular on the use of case studies in programming instruction and on issues arising from teaching LISP in introductory courses. Among the results of their efforts were successful NSF grant proposals, numerous research papers, and two textbooks of case studies. More recently, Clancy has explored "lab-centric" instruction, a technique that swaps lecture and discussion time for supervised hands-on computer lab work. He and his Berkeley colleagues have developed several courses based on this approach. He currently has NSF support to build a community around lab-centric instruction. Results are promising, and research is ongoing.

Christine Cunningham is a vice president at the Museum of Science, Boston, where she oversees curricular materials development, teacher professional development, and research and evaluation efforts related to K-16 engineering and science learning and teaching. Her projects focus on making engineering and science more relevant, understandable, and accessible to everyone, especially marginalized populations such as women, underrepresented minorities, people from low socioeconomic backgrounds, and people with disabilities. She is particularly interested in the ways that the teaching and learning of engineering and science can change to include and also benefit from a more diverse population. Cunningham's projects span the elementary to college educational continuum. Principal among these is Engineering is Elementary (EiE), a project she founded in 2003. EiE is creating a research-driven, standards-based, and classroom-tested curriculum that integrates engineering and technology concepts and skills with elementary science topics. Connections are also made with literacy, social studies, and mathematics. EiE also helps elementary school educators enhance their understanding of engineering concepts and pedagogy through professional development workshops and resources. A research and assessment effort is studying how children and their educators engage with engineering concepts and skills. As the director of EiE, Cunningham is responsible for the vision, strategy, and funding for the project. She received her B.A. and M.A. in biology from Yale University and her Ph.D. in science education and curriculum instruction from Cornell University.

Jan Cuny has been a program officer at the National Science Foundation since 2004, heading the Broadening Participation in Computing Initiative. Before coming to NSF, she was a faculty member in computer science at Purdue University, the University of Massachusetts, and the University of Oregon. Cuny has been involved for many years in efforts to increase the participation of women in computing research. A longtime member of the Computing Research Association's Committee on the Status of Women (CRA-W), she has served among other activities as a CRA-W co-chair, a mentor in its Distributed Mentoring Program, and a lead on its Academic Career Mentoring Workshop, Grad Cohort, and Cohort for Associated Professors projects. She was also a member of the advisory board for the Anita Borg Institute for Women and Technology, the leadership team of the National Center for Women in Technology, and the executive committee of the Coalition to Diversify Computing. She was program chair of the 2004 Grace Hopper Conference and the general chair of the 2006 conference. For her efforts with underserved populations, she is a recipient of one of the 2006 ACM President's Awards and the 2007 CRA A. Nico Habermann Award. Cuny earned a B.A. in computer science from Princeton University, an M.A. from the University of Wisconsin, and a Ph.D. from the University of Michigan, Ann Arbor.

Jill Denner is the associate director of research at Education, Training, Research (ETR) Associates, a non-profit organization in California. She does applied research, with a focus on increasing the number of women and underrepresented minorities in computing. She has developed several after-school programs, and her research on these programs has contributed to an understanding of effective strategies for promoting youth leadership, building youth-adult partnerships, increasing girls' confidence and capacity to produce technology, and engaging girls in information technology. Her current focus is on how students learn while creating computer games, and the development of computational thinking. As part of a long-standing commitment to bridge research and practice, her research is designed and conducted in collaboration with schools and community-based agencies. Denner has been a principal investigator on several NSF grants, written numerous peer-reviewed articles, and co-edited two books: *Beyond Barbie and Mortal Kombat: New Perspectives on Gender and Gaming*, published by MIT Press in 2008, and *Latina Girls: Voices of Adolescent Strength in the US*, published by New York University Press in 2006. She has a Ph.D. in developmental psychology from Teachers College, Columbia University, and a B.A. in psychology from the University of California, Santa Cruz.

Danny C. Edelson is vice president for education at the National Geographic Society and is executive director of the National Geographic Education Foundation. In these positions he leads the National Geographic Society's efforts to improve public understanding of geography and related disciplines through both formal and informal education programs. Throughout his career Edelson has conducted educational research and development, with a primary focus on environmental science and geography. The products of his research and development include My World GIS, a geographic information system (GIS) designed for educational use; *Investigations in Environmental Science: A Case-Based Approach to the Study of Environmental Systems*, a technology-integrated high school environmental science textbook; and Earth science units for two comprehensive, middle school science programs. He has also developed professional development programs for teachers in middle school through college and has led several large-scale instructional reform efforts in urban public schools. Edelson has written extensively on motivation, classroom teaching and learning, educational technology, and teacher professional development, drawing on research conducted with colleagues and students. Prior to joining National Geographic, Edelson was a faculty member in education and computer science at Northwestern University for 13 years, where he founded and directed the Geographic Data in Education (GEODE) Initiative. He is an author on more than 50 papers in journals, edited books, and conference proceedings, including *The Cambridge Handbook of the Learning Sciences, The International Handbook on Science Education, Journal of the Learning Sciences, Journal of Research on Science Teaching*, and *The*

Science Teacher. Edelson received his Ph.D. in computer science (artificial intelligence) from Northwestern University and his B.S. in engineering sciences from Yale University.

Jeri Erickson is the ScienceWorks for ME outreach education coordinator at the Foundation for Blood Research. She is a genetic counselor whose interest in promoting science education and understanding led to her involvement with ScienceWorks for ME, an innovative program designed to offer scientific equipment and professional expertise to Maine's science teachers and their students. She is the project director of both the NIH-funded high school biology curriculum project BiomedicineWorks, and the NSF-funded ITEST project, EcoScienceWorks. She has an M.S. in human genetics from Sarah Lawrence College and a B.A. in biology from Wellesley College.

Louis J. Gross is the James R. Cox Professor of Ecology and Evolutionary Biology and Mathematics and director of the Institute for Environmental Modeling at the University of Tennessee, Knoxville. He is also director of the National Institute for Mathematical and Biological Synthesis, a National Science Foundation-funded center to foster research and education at the interface between math and biology. He has been a faculty member at the University of Tennessee, Knoxville, since 1979. His research focuses on applications of mathematics and computational methods in many areas of ecology, including disease ecology, landscape ecology, spatial control for natural resource management, photosynthetic dynamics, and the development of quantitative curricula for life science undergraduates. He has led the effort at the University of Tennessee, Knoxville, to develop an across-trophic-level modeling framework to assess the biotic impacts of alternative water planning for the Everglades of Florida. He has co-directed several courses and workshops in mathematical ecology at the International Centre for Theoretical Physics in Trieste, Italy, and has served as program chair of the Ecological Society of America, as president of the Society for Mathematical Biology, as president of the the University of Tennessee, Knoxville, Faculty Senate, and as chair of the National Research Council Committee on Education in Biocomplexity Research. He is the 2006 Distinguished Scientist awardee of the American Institute of Biological Sciences and is a fellow of the American Association for the Advancement of Science. He currently serves on the National Research Council Board on Life Sciences and on the board of directors of the American Institute of Biological Sciences. He completed a B.S. in mathematics at Drexel University and a Ph.D. in applied mathematics at Cornell University.

Peter Henderson is co-founder of the math thinking discussion group (www.math-in-cs.org), which advocates the importance of mathematics and mathematical reasoning in computer science and software engineering education. He retired in 2007 as the chair and founder of the Department of Computer Science and Software Engineering at Butler University, and he is currently the editor of two educational columns, "Software Engineering Education" in the ACM Special Interest Group Software Engineering Notes, and "Math CountS" in the ACM Special Interest Group on Computer Science Education InRoads. In addition, he has conducted workshops and given numerous presentations on the role of mathematics in computer science and software engineering education. Henderson has been instrumental in formulating recommendations on the mathematical needs of undergraduate computer science and software engineering programs for the Mathematical Association of America's Committee on the Undergraduate Program in Mathematics, and he has been active at various mathematics and computer science education conferences promoting mathematical thinking. He received his B.S. and M.S. in electrical engineering from Clarkson University. He holds a Ph.D. in electrical engineering from Princeton University and taught computer science and software engineering at SUNY Stony Brook and Butler University from 1974 to 2007.

John R. Jungck is vice president of the International Union of Biological Sciences, president of the IUBS's Commission on Biology Education, and chairperson of the U.S. National Academy of Sciences' National Committee of the IUBS. He is the Mead Chair of the Sciences at Beloit College, the principal investigator (PI) and founder of the BioQUEST Curriculum Consortium, the PI of BEDROCK (Bioinformatics Education Dissemination: Reaching Out, Connecting, and Knitting-together), PI of the SELECTION Working Group of the National Evolutionary Synthesis Center (NESCent), and PI of a subcontract for NUMB3R5 COUNT! (Numerical Undergraduate Mathematical Biology Education). He is the editor of *Biology International* and is on the editorial boards of several journals, including the *Bulletin of Mathematical Biology*, *Evolutionary Bioinformatics*, and *Life Science Education*. Formerly, he was editor of both the *American Biology Teacher* and *Bioscene: Journal of College Biology Teaching*, was president of the Association of College and University Biology Educators, and was chairperson of the Education Committee of the Society for Mathematical Biology for 14 years. He serves on numerous boards such as for the National Electronics and Computer Technology Center (NECTEC) in Thailand, the Alan C. Wilson Center for Molecular Evolution in New Zealand, and the National Institute for Mathematical Biology Synthesis Center (NIMBioS) in the United States, and he is on the revision committee of the College Board

Advanced Placement Biology program. He is an international leader in biology education reform, a mathematical molecular evolutionary biologist, and a computer software developer of biological simulations, tools, and databases. His research interests include the origins of genetic codes, patterns in nature, and evolutionary analysis of complex data sets. His awards, honors, and offices include AAAS Fellow, an honorary doctorate from the University of Minnesota, American Institute of Biological Sciences Education Award, Fulbright Scholar to Chiang Mai University in Thailand, Mina Shaughnessy Scholar of the U.S. Department of Education, and a National Science Teachers Association Gustav Ohaus Award for Outstanding Innovations in College Science Teaching. Jungck earned a B.S. in biochemistry and an M.S. in genetics and microbiology from the University of Minnesota. He received his Ph.D. in evolution from the University of Miami.

Deanna Kuhn is a professor of psychology and education at Teachers College, Columbia University. She was previously a faculty member at the Harvard Graduate School of Education. She has published widely in psychology and education, in journals ranging from *Psychological Review* to *Harvard Educational Review*. She has written three major books: *The Development of Scientific Thinking Skills*, *The Skills of Argument*, and, most recently, *Education for Thinking* (Harvard University Press, 2005). She is editor of the journal *Cognitive Development*, a former editor of the journal *Human Development*, and co-editor of the last two editions of the *Cognition* volume of the *Handbook of Child Psychology*. In recent years, her work has focused on developing inquiry and argument curricula for middle schoolers. Her Ph.D. is from the University of California, Berkeley, in developmental psychology.

Cathy Lachapelle, director of research and evaluations for the Engineering is Elementary (EiE) project at the Museum of Science, currently leads the assessment and evaluation efforts for the EiE curriculum, designing assessment instruments, piloting and field-testing them, and conducting research on how children use the EiE materials. She has worked on a number of research and evaluation projects related to K-16 engineering education. She has worked in numerous classrooms studying children's learning of science, mathematics, and engineering content and processes. She received her S.B. in cognitive science from MIT and her Ph.D. in psychological studies in education from Stanford University.

Joyce Malyn-Smith is strategic director of the Workforce and Human Development Program for the Education Development Center, Inc.'s Learning and Teaching Division. Her unique combination of experience includes more than 10,000 contact hours as a K-12 classroom teacher and 13 years as a public school administrator responsible for curriculum and professional development in more than 30 career and technical education programs in 15 high schools. Her EDC projects help expert workers articulate their skills and knowledge and develop systems and tools to integrate these into programs and curricula for K-20. With ongoing interests in youth who are power users of technology, Malyn-Smith is a PI for NSF's ITEST Learning Resource Center, serving more than 160 ITEST projects, and leads its working group on computational thinking. She is also a PI for the NSF-ATE IT Across Careers (I-III). She led the U.S. Education Department's IT Career Cluster Initiative in creating the national career cluster model and curricular framework used in 49 states. She developed national voluntary skill standards for bioscience, human services, and chemical process industries and co-authored *Making Skill Standards Work* (U.S. Department of Labor). She led the development of scenario-based assessments for New York and rubrics for NSF's ITAC projects and the U.S. Department of Energy's Real World Design Challenge (2008-2010). Malyn-Smith served on the ETS International ICT Literacy Panel (*Digital Transformation: A Framework for ICT Literacy*). She currently serves on Certiport's Global Digital Literacy Council and other project advisory boards. A USOE Fellow in Bilingual Education, Malyn-Smith holds a doctorate from Boston University, a master's degree from Boston State Teacher's College, and a B.S. from Universidad InterAmericana, Puerto Rico.

Taylor Martin joined the faculty at the University of Texas at Austin in 2003. Her primary research interest is how people learn content in complex domains from active participation, both physical and social. She is cooperating with local elementary schools to improve tools for assessing young children's learning of mathematics and to examine how learning is affected by hands-on activities, and she is investigating the development of adaptive expertise through cooperation with the Vanderbilt–Northwestern–Texas–Harvard (VaNTH)/MIT Engineering Research Center in Bioengineering Educational Technologies. Martin received a B.A. in linguistics and an initial teaching certification from Dartmouth College in 1992, an M.S. in psychology from Vanderbilt University in 2000, and a Ph.D. in education from Stanford University in 2003.

Robert Panoff is founder and executive director of the Shodor Eduation Foundation, a non-profit education and research corporation dedicated to the reform and improvement of mathematics and science education through computational and communication technologies. As PI on several National Science Foundation (NSF) and U.S. Department of Education grants that explore interactions between technology and education, he develops interactive simulation modules that combine standards, curriculum, supercomputing resources, and desktop computers. In recognition of Panoff's efforts in college faculty enhancement and curriculum development, the Shodor Education Foundation was named as an NSF Foundation Partner for the revitalization of undergraduate education. Shodor established the Shodor Computational Science Institute, which was expanded with NSF funding to become the National Computational Science Institute. Shodor's Computational Science Education Reference Desk was funded as a Pathway portal of the National Science Digital Library. Panoff consults at several national laboratories and is a frequent presenter at NSF workshops on visualization, supercomputing, and networking. He has served on the NSF advisory panel for the Applications of Advanced Technology program, and he is a founding partner of the NSF-affiliated Corporate and Foundation Alliance. Panoff received his M.A. and Ph.D. degrees in theoretical physics from Washington University in St. Louis, with both pre- and postdoctoral work at the Courant Institute of Mathematical Sciences at New York University. Wofford College awarded Panoff an honorary doctor of science degree in recognition of his leadership in computational science education.

Mitch Resnick, a professor of learning research at the MIT Media Lab, develops new technologies to engage people (especially children) in creative learning experiences. His Lifelong Kindergarten research group developed the "programmable bricks" that were the basis for the LEGO Mindstorms robotics kits, and he co-founded the Computer Clubhouse network of after-school learning centers for youth from low-income communities. Resnick's group recently developed a programming language and online community called Scratch (http://scratch.mit.edu), which enables children to create their own interactive stories, games, animations, and simulations—and share their creations online. In the process, children learn to think creatively, reason systematically, and work collaboratively. Resnick earned a B.S. in physics from Princeton University and an M.S. and a Ph.D. in computer science from MIT. He worked for 5 years as a science and technology journalist for *Business Week* magazine, and he has consulted around the world on the uses of new technologies in education. He is the author or co-author of several books, including *Turtles, Termites, and Traffic Jams*.

Christina Schwarz is an associate professor of teacher education at Michigan State University. Her research centers on teaching and learning science and the role that technology might play in this process. She focuses specifically on inquiry-oriented and model-centered constructivist learning environments, particularly at the elementary and middle school levels. Her current research involves helping students and teachers develop an understanding of scientific practices such as inquiry and modeling and helping them learn how to engage in those practices. Other interests include teacher and student learning progressions, frameworks for teaching science, educational technology, science teaching and learning in urban schools, science curriculum development and evaluation, and the history and philosophy of science. Schwarz received her Ph.D. from the University of California, Berkeley.

Jim Slotta, a professor at the University of Toronto's Ontario Institute for Studies in Education, teaches a graduate-level course titled "Special Topics: Doctoral Level: Technology, Cognition and Instruction." He has been involved in a group research project to encourage school district partners to use technology in classrooms. He is a recipient of the IBM Faculty Award for e-Learning Design, 2003. Slotta has received a variety of grants from public and private organizations. Currently, he is co-principal investigator of several funded research projects, including a 3-year NSF-funded project titled "Partnership Model for Integrating Technology, Curriculum, and Professional Development in Response to New Science Assessments," a 5-year NSF-funded project titled "The Educational Accelerator Center: Technology-Enhanced Learning in Science (TELS)," and a 2-year German DFC-funded research project titled "NetCoIL: Scientific Network for Collaborative Inquiry Learning."

Matthew Stone is an associate professor in the Department of Computer Science and Center for Cognitive Science at Rutgers University. He received his Ph.D. in 1998 in computer and information science from the University of Pennsylvania. He was a postdoctoral fellow at Rutgers from 1998 to 1999 and joined the faculty in 1999. From 2005 to 2006 he was a visiting fellow in the School of Informatics at the University of Edinburgh. He serves on the editorial board of the journals *Computational Linguistics* and *Artificial Intelligence* and just served as program co-chair for the 2007 North American Association for Computational Linguistics Human Language Technology Conference. His research is funded by the NSF.

Robert Tinker has, for 30 years, pioneered research on innovative approaches to education that exploit technology. The initial development of probeware for learning based on real-time measurements was performed in his group. His team at TERC was the first to develop "network science" for dispersed science investigations. The initial result of this work was the NGS Kids Network, a groundbreaking curriculum that was the first to make extensive use of student collaboration and data sharing. Fifteen years ago he founded the non-profit Concord Consortium to concentrate on innovative applications of technology in education. The Concord Consortium developed some of the earliest professional development based on online courses. This led to the establishment of the Virtual High School, a fully accredited high school that has a unique low-cost cooperative design. Current work focuses on sophisticated simulations in science, probeware, and handhelds, and applications of these technologies to pressing educational issues, with a particular focus on underrepresented students. A current focus is applying technology to monitoring student progress and to supporting diverse learners. The open-source, free technologies emerging from the Concord Consortium are being integrated into learning modules that offer a glimpse of what inquiry-based education could look like in a few years. Tinker earned his Ph.D. in experimental low-temperature physics from MIT and has taught college physics for 10 years. His focus on education developed as a result of teaching at Stillman College, a historically African American college in Alabama.

Stephen Uzzo is vice president of technology at the New York Hall of Science, where he focuses on a number of projects related to science, technology, engineering, and mathematics (STEM) learning; sustainability; and network science, including "Connections: the Nature of Networks," a public exhibition on network science that opened in 2004. He was also the local organizer for the 2007 International Conference and Workshop on Network Science. In addition to his work at the Hall of Science, Uzzo serves on the faculty of the New York Institute of Technology Graduate School of Education, where he teaches STEM teaching and learning. During the 1980s, he worked on a number of media and technology projects. In 1981, he was part of the launch team for MTV and was appointed chief engineer for video/computer graphics production and distance learning networks for the NYIT Video Center in 1984. Other projects during that period included the first all-digital satellite television transmission, best practices group for the NBC Summer Olympic Games in Barcelona, and a team of scientists and engineers at the Space Studies Institute at Princeton to develop and test lunar teleoperations simulators. During the 1990s, Uzzo served on numerous advisory boards for educational institutions, as well as facilitating major technology initiatives among K-12 public/private schools, higher education, and government to improve STEM literacy. His work on various projects important to conservation includes ecosystems studies that were instrumental in blocking offshore oil drilling in New York waters and a cross-sound bridge in Oyster Bay, as well as cleanup planning for Superfund sites. He has worked on preservation and open space projects on Long Island and the San Francisco Bay Peninsula. He holds a Ph.D. in network theory and environmental studies from the Union Institute.

Michelle Williams is an assistant professor of science education in the Department of Teacher Education at Michigan State University. Her research focuses on both teaching and learning in science and technology education. She recently received a National Science Foundation Faculty Early Career Development Award titled "CAREER: Tracing Children's Developing Understanding of Heredity over Time." Williams' current work explores how upper elementary and middle school students develop coherent understandings of genetic inheritance and related ideas within and across successive grades using the Web-based Inquiry Science Environment. She earned a B.B.A. in marketing from the University of Texas at Austin in 1992 and an M.A. and a Ph.D. in development in mathematics and science from the University of California, Berkeley, in 2001 and 2004.

Jeannette Wing is the President's Professor of Computer Science in the Computer Science Department at Carnegie Mellon University. From 2004 to 2007, she was head of the Computer Science Department at Carnegie Mellon University. From 2007 to 2010 she was the assistant director of the Computer and Information Science and Engineering Directorate at the National Science Foundation. Wing's general research interests are in the areas of specification and verification, concurrent and distributed systems, programming languages, and software engineering. Her current focus is on the foundations of trustworthy computing, with specific interests in security and privacy. She published a viewpoint article in the March 2006 issue of *Communications of the Association for Computing Machinery* entitled "Computational Thinking." Wing received her S.B., S.M., and Ph.D. from the Massachusetts Institute of Technology.

Ursula Wolz is an associate professor of computer science and interactive multimedia at the College of New Jersey, is the principal investigator for the NSF program Broadening Participation in Computing via Community Journalism for Middle Schoolers, and was the principal investigator for a Microsoft Research project on multidisciplinary game development. She is a recognized computer science educator with a broad range of publications who has taught students including disabled children, urban teachers, and elite undergraduates for more than 30 years. She is a co-founder of the Interactive Multimedia Program at the College of New Jersey. She has a background in computational linguistics, with a Ph.D. in computer science from Columbia University, a master's degree in computing in education from Columbia Teachers College, and a bachelor's degree from MIT, where she was part of Seymour Papert's Logo group at the very beginning of research on constructivist computing environments.

B.3 STAFF

Herbert S. Lin, the study director, is chief scientist for the National Research Council's Computer Science and Telecommunications Board, where he has been a study director for major projects on public policy and information technology. These studies include a 1996 study on national cryptography policy (*Cryptography's Role in Securing the Information Society*), a 1991 study on the future of computer science (*Computing the Future*), a 1999 study of Defense Department systems for command, control, communications, computing, and intelligence (*Realizing the Potential of C4I: Fundamental Challenges*), a 2000 study on workforce issues in high technology (*Building a Workforce for the Information Economy*), a 2002 study on protecting kids from Internet pornography and sexual exploitation (*Youth, Pornography, and the Internet*), a 2004 study on aspects of the FBI's information technology modernization program (*A Review of the FBI's Trilogy IT Modernization Program*), a 2005 study on electronic voting (*Asking the Right Questions About Electronic Voting*), a 2005 study on computational biology (*Catalyzing Inquiry at the Interface of Computing and Biology*), a 2007 study on privacy and information technology (*Engaging Privacy and Information Technology in a Digital Age*), a 2007 study on cybersecurity research (*Toward a Safer and More Secure Cyberspace*), a 2009 study on health care information technology (*Computational Technology for Effective Health Care*), and a 2009 study on cyberattack (*Technology, Policy, Law, and Ethics Regarding U.S. Acquisition and Use of Cyberattack Capabilities*). Before his NRC service, he was a professional staff member and staff scientist for the House Armed Services Committee (1986-1990), where his portfolio included defense policy and arms control issues. He received his doctorate in physics from MIT.

Enita A. Williams is an associate program officer with the Computer Science and Telecommunications Board of the National Research Council. She formerly served as a research associate for the NRC's Air Force Studies Board, where she supported a number of projects, including those of a standing committee for the Special Operations Command (SOCOM) and a standing committee for the intelligence community (TIGER). Prior to her work at the NRC, she served as a program assistant with the Scientific Freedom, Responsibility and Law Program of the American Association for the Advancement of Science, where she drafted the human enhancement workshop report. Ms. Williams graduated from Stanford University with a B.A. in public policy with a focus on science and technology policy and an M.A. in communications. She is currently pursuing a law degree at Georgetown University Law Center.

Shenae Bradley is a senior program assistant at the Computer Science and Telecommunications Board of the National Research Council. She currently provides support for the Committee on Sustaining Growth in Computing Performance, and has worked with the Committee on Wireless Technology Prospects and Policy Options, among others. She formerly served as an administrative assistant for the Ironworker Management Progressive Action Cooperative Trust and managed a number of apartment rental communities for Edgewood Management Corporation in the Maryland/DC/Delaware metropolitan areas. She is in the process of earning her B.S. in family studies from the University of Maryland at College Park.